THE CHEMISTRY
OF CERAMICS

THE CHEMISTRY
OF CERAMICS

Hiroaki Yanagida
University of Tokyo, Japan

Kunihiko Koumoto
Nagoya University, Japan

Masaru Miyayama
University of Tokyo, Japan

Translated by
Hisao Yamada
Cerone, Inc., Ohio, USA

JOHN WILEY & SONS

CHICHESTER · NEW YORK · BRISBANE · TORONTO · SINGAPORE

Authorised Translation from Japanese
language edition published by
Maruzen Co., Ltd, Tokyo,
Hiroaki Yanagida, Emeritus Prof., The University of Tokyo
Kunihiko Koumoto, Prof., School of Engineering, Nagoya University
Masaru Miyayama, Associate Prof., RCAST, The University of Tokyo

Originally published by Maruzen Co., Ltd, Tokyo

Other Wiley Editorial Offices

John Wiley & Sons, Inc., 605 Third Avenue,
New York, NY 10158-0012, USA

Jacaranda Wiley Ltd, 33 Park Road, Milton,
Queensland 4064, Australia

John Wiley & Sons (Canada) Ltd, 22 Worcester Road,
Rexdale, Ontario M9W 1L1, Canada

John Wiley & Sons (Asia) Pte Ltd, 2 Clementi Loop #02-01,
Jin Xing Distripark, Singapore 129809

Library of Congress Cataloging-in-Publication Data

British Library Cataloguing in Publication Data

A catalogue record for this book is available from the British Library

ISBN 0 471 95627 9 (Hbk)
ISBN 0 471 96733 5 (Pbk)

Typeset in 10/12pt Times by Dobbie Typesetting Ltd, Tavistock, Devon
Printed and bound in Great Britain by Biddles, Guildford, Surrey
This book is printed on acid-free paper responsibly manufactured from sustainable forestation,
for which at least two trees are planted for each one used for paper production.

CONTENTS

PREFACE TO THE ENGLISH VERSION

Since 'The Chemistry of Ceramics' was first published by Maruzen in 1982, I am delighted to say that it has been adopted as a textbook for courses at many universities and technical colleges. This would seem to indicate that the authors correctly recognised a need for such a book and produced one that fulfilled lecturers needs exactly. The book is written primarily for students with a chemistry background with less descriptive detail than a traditional textbook on ceramic materials and products and a different emphasis than would be required by students of physics. The authors' intentions have been to present their perspective on ceramic materials assuming the reader has an understanding of the basic science involved.

The use of ceramics goes back many thousands of years, however until recently, the history of ceramics and knowledge gained had not been systematically documented. The first edition of this book was an important step towards this.

The advanced ceramics industry has progressed greatly since then. For example; Information Technology is now based upon a variety of materials, amongst which electric ceramics play an important role; the precision machining industry could not exist without high performance machinery ceramics and many environmental monitors are now made of ceramics, including automobile exhaust sensors. Ceramics are also proving to be very promising materials for generating clean energy.

Along with the development of the advanced ceramics industry, scientific knowledge of the materials has been accumulated and systemised. Since the first edition, I have been trying to revise the content and the first major opportunity I had to do this was when Maruzen decided to publish a second edition in 1993. This English version is primarily based on the second edition and the content has been thoroughly checked and revised where necessary during the translation process. I am proud to say that the second edition was also well received and is now thought of as *the* standard textbook for ceramics.

The science of ceramics and the ceramics industry in Japan have developed and improved greatly over the last twenty years. This book is regularly consulted by many people working in the field especially when trying to gain a clearer scientific perspective on ceramics.

I am very grateful to my publishers, Maruzen and John Wiley & Sons for giving me the opportunity to edit the English version and would also like to thank Hisao Yamada for the excellent job he has done in translating the book. It is a remarkable coincidence that I am now based in London for much of the time in my position as the first director of the Japan Society for the Promotion of Science, as well as retaining my principal job as Professor of Ceramics and intelligent materials at the University of Tokyo. I hope and believe that this English version will help the development of ceramic science and industry worldwide.

Mid-December 1995

Hiroaki Yanagida
Principal Author

PREFACE TO THE FIRST EDITION

The progress of technology has been supported by the development of new materials. Ceramics, which makes up the three major classes of materials along with metals and plastics, have made remarkable progress since the early 1970s. Progress in integrated circuits has been supported by the development of alumina substrates. Optical fibers for telecommunications have been commercialized since the establishment of manufacturing processes for high-purity glass fibers. The performance of cutting tools has been improved by the introduction of hard ceramics such as diamonds. Ceramics with superior electromagnetic, optical and chemical characteristics along with conventional ceramic characteristics such as hardness, refractoriness and anti-corrosion properties have emerged as a group of high-precision ceramics which are now called fine ceramics. It is widely anticipated that the science and technology of ceramics will develop further in the future. Ceramics will be applied widely in combustion engines, energy development, medical implants and microelectronics.

In order to understand ceramics in detail and to provide a stepping stone for further development, it is necessary to have a concise textbook which covers a wide variety of subjects from basics to applications. Ceramics textbooks in the market today are either too obsolete without any coverage of new ceramics or too voluminous due to the inclusion of a wide variety of subjects.

In order to overcome these shortcomings in existing textbooks, attempts have been made in this book to cover the minimal number of essential concepts for understanding ceramics. Thus the book employs a group of terminology for chemistry and is intended for young chemistry students. This book is based on my lectures on 'Solid State Chemistry' at Tokyo University. Fifteen lectures, each an hour and a half long, were given to undergraduate students whose major was Applied Chemistry. The lectures have now been expanded to solid structures and solid reactions and are given in 22 to 23 sessions. The manuscript has been assisted by Drs Kunihiko Koumoto (now Professor at Nagoya University) and Masaru Miyayama, two associate professors at University of Tokyo, who audited my lectures. Because of the involvement of these researchers I believe that the book has been written consistently from the viewpoint of chemistry. I would like to take this opportunity to thank both of them.

It is my pleasure to see that this book contributes in furthering the interests of ceramics, provides a basic understanding in ceramics from chemical viewpoints and becomes a stepping stone for the further development of ceramics.

Finally I would like to acknowledge the assistance of the staff at Maruzen Publishing Company for the publication of this book.

Early spring, 1982

Hiroaki Yanagida
Author/Editor

PREFACE TO THE SECOND EDITION

More than ten years have passed since the first publication of this book in the applied chemistry series. During the period ceramics have made remarkable progress both scientifically and technically. Some of the examples are as follows. Ceramic turbochargers have been commercialized as components in automobile engines. A group of ceramics has been discovered to exhibit superconductivity at high temperatures, which was unthinkable ten years ago. Also, piezoelectric motors without a coil have been introduced commercially.

This book was written originally to cover the chemistry of ceramics from basics to applications so that readers would obtain an understanding of the physical properties of ceramics. In light of the remarkable progress in ceramics over the last ten years it occurred quite naturally to the authors to revise the book by incorporating some new and interesting topics.

The authors have also gained invaluable experience during the last ten years. The editor (Yanagida) played a leading role in the establishment of the Research Center for Advanced Technologies in May 1987 at Tokyo University and was the director of the Center from April 1989 to March 1991. During the period he led educational and research activities at the Center with academic excellence, internationality, openness and mobility as his motto. He felt strongly about the needs of research as well as development activities, and he began with these ideals. One of the co-authors, Dr Kunihito Koumoto who was an associate professor ten years ago, has been appointed a professor at the Faculty of Engineering at Nagoya University, a Mecca of ceramics in Japan. Dr Masaru Miyayama, who was a senior research associate ten years ago, participated in the establishment of the Research Center for Advanced Technologies as an associate professor from its inception. He is playing a leading role at the Center today. We have made every effort to reflect our experience in revising the book. A wide acceptance of the first edition made it possible to obtain the cooperation of the Maruzen Publishing Company for a revision.

It is hoped that this book will contribute to the further development of the science and technology of ceramics.

Mid-autumn 1993   Hiroaki Yanagida,
                                                             Author/Editor

1

INTRODUCTION

1.1 THE POSITION OF CERAMICS IN MATERIALS SCIENCE

Ceramics have been defined as non-metallic, inorganic solid materials produced by thermal treatment. Compared to metals and plastics, ceramics are hard, non-combustible and non-oxidizable. Thus they can be used in severe high-temperature, corrosive and tribological environments. In addition, many ceramics exhibit superior electromagnetic, optical and mechanical properties under these environments. Because of these unique characteristics, ceramics have been increasingly sought for energy development and advanced telecommunications.

The term 'material' is the most important aspect of the above definition of ceramics. Without this term, even volcanic ash and lava can be regarded as ceramics. If it is not possible to fabricate shapes with given dimensions, it is no use having a new inorganic substance with superior properties. Although the gap between a substance and a material is quite small in metals and plastics (due to their easier fabrication), the difference is significant in ceramics.

Ceramic materials which overcame this gap in the past include porcelains, glasses and cements. Porcelains are made of three essential ingredients, namely, siliceous stone as a skeletal ingredient, clay as a forming additive and feldspar as a sintering additive. These three ingredients have three essential functions, namely, plasticity of aqueous slurry from clay, dry strength, and bonding of siliceous stones by the fusion of feldspar during sintering. In order to manufacture porcelains into desired shapes and to provide functionality it is essential to have these three materials. Siliceous stone, clay and feldspar are all silicate minerals.

Glasses soften when heated and it becomes possible to fabricate them into shapes. When cooled to room temperature, glasses retain their shape at high temperatures and thus become materials. The major ingredients of glasses are soda ash and a silicate which contains some calcium oxide. When mixed with water, cements become plastic and it becomes possible to fabricate shapes. During subsequent hydration, an aqueous mixture of a cement coagulates and solidifies into a desired shape. Calcium silicates are the major ingredients of cements. Since porcelains, glasses and cements all contain silicates, the manufacture of these ceramics is classified as the silicate industry. The main functionality of these ceramics is either structural or containment.

Silicates are not the only non-metallic, inorganic solid materials obtained by thermal treatment. Non-oxides such as nitrides and carbides and composites with metals and plastics exhibit superior properties which cannot be obtained from silicates. It is vitally important to develop manufacturing processes so that the superior functions of these substances can be fully exploited. When developed, it is possible to produce ceramic materials with superior electromagnetic, optical and mechanical functions together with the more conventional characteristics of hardness, non-combustibility and anti-corrosiveness. Manufacturing technologies of these substances have improved significantly during the last few decades and many non-metallic inorganic solids have been able to be transformed into ceramics. These new classes of ceramics are now called either new or fine ceramics. Table 1.1 lists a number of new ceramics. The distinction between new and fine ceramics can be made as follows. Those non-silicate ceramics which can be exploited for a few of their unique characteristics are called new ceramics. When most of their unique characteristics can be exploited, ceramics are called fine ceramics. Table 1.2 lists new and fine ceramic applications of alumina (Al_2O_3).

The progress of conventional ceramics to fine ceramics is summarized in Table 1.3. Characteristics of fine ceramics are compared in Table 1.4 with those of stoneware, metals and plastics. As indicated in Table 1.4, brittleness is the most significant deficiency of fine ceramics. This deficiency needs to be overcome in future.

1.2 CERAMICS IN THE HISTORY OF MATERIALS

The very first tools that human beings used were stone (e.g. spades and axes) but it was not possible to make any large and complicated shapes. Later a technique was developed to make earthenware by mixing clay with water, shaping and firing. The earthenware was used as containers, but was prone to leakage. Thus stoneware and earthenware were the precursors of ceramics. Although the stoneware was hard and resistant to both heat and water, it was very difficult to shape and fabricate. On the other hand, earthenware was easy to work but it was weak and not watertight. Therefore, in order to improve both stoneware and earthenware it was necessary to improve fabrication of the stoneware and the physical properties of the earthenware.

Bronze was the first metallic material and was used extensively to produce a variety of containers and cutting tools because of its excellent ductility. But

Table 1.1 Functions and applications of fine ceramics

Functions	Substances and states	Elements	Devices
(A) *Electrical and magnetic*			
A-1: high insulation	Al_2O_3 (HP/HD SB, SX plates)	IC substrates	ICs
	AlN (HP/HD SB)	Heat spreaders	ICs
	C (HP SX)	Heat spreaders	ICs
A-2: dielectric	$BaTiO_3$ (HP/HD SB, SX)	HC capacitors	Electronic circuits
	$Bi_2O_3 \cdot SnO_2$ (HP/HD SB)	HC capacitors	Electronic circuits
A-3: piezoelectric	$Pb(Zr_xTi_{1-x})O_3$ (polarized HD SB)	Oscillators, ignitors, filters, transformers	Ultrasonic elements, electric circuits
	ZnO (oriented thin films)	Surface elastic wave delay elements	Electronic circuits
A-4: pyroelectric	SiO_2 (thin film SX)	Oscillators	Watches
	$Pb(Zr_zTi_{1-z})O_3$ (polarized HD SB)	IR detectors	
A-5: strong dielectric	$(1-x)Pb(Zr_zTi_{1-z})O_3 + xLa_2O_3$ (translucent HD SB)	Image storage elements, optoelectronic deflectors	
A-6: soft magnetic	$Zn_{1-x}Mn_xFe_2O_4$ (HD SB, GB controlled)	Memory, calculators, magnetic cores, magnetic tapes	Computers, transformers, tape recorders
	γ-Fe_2O_3 (needle-shaped fine powder)		
A-7: hard magnetic	$SrO \cdot 6Fe_2O_3$ (oriented HD SB)	Magnets	CRT
	$SrO \cdot 6Fe_2O_3$ (powder/rubber composites)	Plastic magnets	Airtight shutters for refrigerators
A-8: semiconducting	$La_{1-x}Ca_xCrO_3$ (SB)	Resistance heaters, gas sensors, thermistors	HT furnaces, gas-leak detectors, temp. controllers, lightning arresters
	SnO_2 (porous SB, Pt loaded) transition metal oxides (HD SB)		
	ZnO-Bi_2O_3 (MS-controlled SB)	Varistors	Surge arresters
	$BaTiO_3$ (MS-controlled SB)	Self-controlled resistance heaters	Electronic jars, bed dryers

(*continued*)

Table 1.1 (*continued*)

Functions	Substances and states	Elements	Devices
A-9: ionic conduction	β-Al$_2$O$_3$ (HD SB) stabilized zirconia (HD SB)	Na-S batteries, oxygen sensors	Load leveling for power generators; Blast furnace control, air/fuel controllers for automobiles
A-10: electron emission	LaB$_6$	Thermal cathodes for electron guns	Thermal electron devices
A-11: superconductor	Ba$_2$LaCu$_3$O$_{7-\delta}$	Josephson junctions; Superconducting magnets	High-speed computers; Magnetic levitation
(B) High hardness			
B-1: polishing, grinding, cutting	Al$_2$O$_3$, B$_4$C, diamond (powders); Al$_2$O$_3$, B$_4$C, diamond (resin bonded); Al$_2$O$_3$, B$_4$C, diamond (metal bonded); TiN, TiC, B$_4$C, Al$_2$O$_3$ (HD SB)	Polishing, grinding stones; Cutting tools; Cutting tools	
B-2: mechanical strength	Si$_3$N$_4$, SiC (HD SB)	Turbine blades	Automobile engines
(C) Optical			
C-1: fluorescence	Y$_2$O$_3$: Eu (powders)	Fluorescence	Color televisions
C-2: translucence	Al$_2$O$_3$ (translucent SB); SnO$_2$ (coating layers)	HT, CR translucent bodies; Visible light-transmitting semiconductors	High-pressure sodium lamps; Antifog windows
C-3: optical deflection	PLZT (see A-5 above)	HT metallic characteristics	Solar collectors
C-4: optical reflection	TiN (thin film coating)	Visible light transmission with IR reflection	Energy-saving window panes
C-5: IR reflection	SnO$_2$ (thick film coating)		
C-6: optical transmission	SiO$_2$ (HP fibers)	Optical fibers	Optical transmission cables, optical fiber cameras

(D) Thermal			
D-1: thermal stability	ThO_2 (HD SB)	Thermally insulating structural components	HT furnaces
D-2: thermal insulation	$K_2O \cdot nTiO_2$ (fibers) $CaO \cdot nSiO_2$ (foams)	Thermally stable insulators Lightweight insulators	Energy-saving furnaces Non-combustible insulators
D-3: thermal conduction (see A-1 above)	AlN (HP, HD SB) C (HP SX)	Electronic substrates Electronic substrates	ICs ICs
(E) Chemical and biochemical			
E-1: artificial bones	Al_2O_3, $Ca_5 (F, Cl) P_3O_{12}$ (HP SB)	Artificial bones and teeth	Bioceramics
E-2: substrates	SiO_2 (porous bodies with controlled pore sizes)	Substrates for enzymes	Biochemical applications
E-3: catalytic	Al_2O_3, TiO_2 (porous bodies) $K_2O \cdot nAl_2O_3$ (porous SB)	Substrates for catalysts Catalysts for aqueous gas reactions	Chemical reaction control Catalytic applications at HT

CR: corrosion resistant.
HC: high capacity.
HP: high purity.
IC: integrated circuit.
MS: microstructure.
SX: single crystal.
CRT: cathode ray tube.
HD: high density.
HT: high temperature.
GB: grain boundary.
SB: sintered body.

Table 1.2 Transition of alumina to fine ceramics

Superior characteristics of alumina	Cutting tools	IC substrates	Discharge tubes for sodium lamps
Thermal stability	Δ	O	⊕
High hardness	⊕	O	O
Heat conduction	Δ	⊕	⊕
Electrical insulation	x	⊕	⊕
Chemical stability	Δ	O	⊕
Optical translucency	Δ	Δ	⊕
Notes (additives)	The characteristics marked by x has been degraded by sintering additives.	Some substrates are sintered with a small amount of sintering additives. The presence of radioactive substances is a problem.	A small amount of MgO has been added to control grain growth. It is necessary to avoid the segregation of the additive.

⊕ denotes the unique characterstics which makes alumina a material.
O denotes additional characteristics of alumina.
Δ denotes the characteristics which require a major improvement.
x denotes the characteristics which are not exploited or not used.

Table 1.3 The progress of ceramics

Class of ceramics	Applications	Substances and characteristics
Conventional ceramics	Glasses, cements and porcelains	Silicates (SiO_2 as a main ingredient)
New ceramics	Refractory, polishing and cutting materials	Materials other than silicates
Fine ceramics	Electronic materials and precision machining tools	Compositional and microstructural control

Table 1.4 The position of ceramics in materials science

Class of materials	Merits	Demerits
Stoneware	Hard	Poor fabricability
Iron (metal) polymers	Ductility	Rusting
	Easy to shape and fabricate	Combustible
Fine ceramics	Have advanced characteristics (hard, non-combustible and corrosion resistant)	Brittle

bronze is not hard or resistant to corrosion. Moreover, it is weak at elevated temperatures. Thus this material was replaced subsequently by steel, a superior material to bronze.

The development of steel paved the way for industrialization. But steel is also prone to heat and corrosion and it was not possible to utilize steel under severe environments. This made necessary the development of high-temperature alloys and stainless steels.

The advent of the petrochemical industry made possible the mass production and wide use of plastics. The superior properties of plastics enabled the fabrication of a large variety of products such as containers, fibers and membranes. It is not an exaggeration to say that the high-growth period of the petrochemical industry corresponded to the development of plastics. The greatest deficiency of plastics is their inability to withstand heat. Although some heat-resistant plastics have been developed, they are far inferior in this respect to either metals or ceramics.

Let us look at the progress of both stoneware and earthenware. The former has made little progress since the Stone Age. The only stone tools still in use today are those used for grinding. It was not possible to machine any hard ceramics until a few decades ago when diamond tools were introduced extensively. Until then the machining of hard materials was mainly in the manufacture of jewelry. On the other hand, it became possible to improve some physical characteristics of earthenware while maintaining its good manufacturing properties. The development of glazes solved the problem of their permeability to water. Glazes also marked the beginning of the glass industry. With increasing firing temperatures earthenware became denser and mechanically stronger. Porcelain is one of the most common forms of earthenware. Cements are the other means to manufacture ceramics. When mixed with water, calcia and plaster become plastic, which makes it possible to form shapes. However, calcia and plaster have poor water resistance. The appearance of water-resistant cements had to wait until the beginning of the nineteenth century when Portland cement became available. Cements are very similar to earthenware and consist also of silicates. Because they are amenable to shaping and manufacturing of forms, silicates were the first non-metallic inorganic materials to be industrialized. It is interesting to note here that most new and fine ceramics under development today are neither silicates not oxides. Compared to silicates they are superior in thermal, chemical, tribological and electrical properties.

The key to manufacturing many materials is the ability to form desired shapes and to sinter without sacrificing the materials' unique characteristics (an extension of earthenware-porcelain technology) and to machine hard ceramics (an extension of stoneware technology). The wide variety of functionality of new ceramics is listed in Table 1.1. Thus it is fair to say that from a viewpoint of materials science the knowledge-intensive age today can be called either the

Second Stone Age or the Artificial Stone Age (or Ceramic Age). This progress is shown in Fig. 1.1.

1.3   THE CHEMISTRY OF CERAMICS AS AN ACADEMIC TOPIC

The number of substances which are regarded as ceramics is increasing very rapidly. Every substance has its unique physical properties, a unique structure and a unique method of manufacture. Thus academic study of ceramics has been expanding very quickly. Physical properties, structure and manufacturing methods (reactions) are the three areas of materials research. It is no use studying only one aspect. For example, without considering the origin of a unique physical property, study of that property is merely measurement of it. Without learning how to manufacture it is no use investigating the structure of a substance with a unique physical property simply because the substance cannot be used as a material. The 'trial and error' synthesis of a substance will not attract much interest without identifying its structure and physical properties.

One of the major characteristics of ceramics is their structural diversity. In general, 'structure' implies a crystal structure, but in ceramics a structure means one of the following:

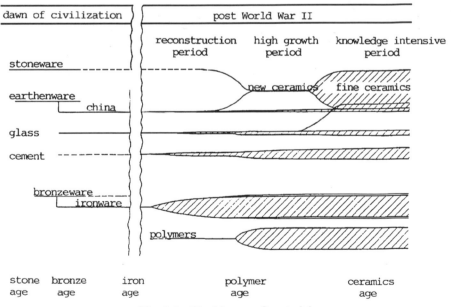

Fig. 1.1   The history of materials

(1) An electron orbit
(2) An atomic coordination
(3) A crystal chemical structure including the second-nearest neighbors in glasses
(4) A crystal structure
(5) Lattice defects
(6) Dislocations
(7) Stacking faults including twins
(8) A surface structure
(9) An interfacial or joint structure
(10) A void structure (radius and its distribution)
(11) A grain structure (grain size and its distribution, density, preferred orientation and morphology).

These structures are closely interrelated with each other. Thus it is necessary to grasp the diversity and interrelationship of the structures not only to analyze the structures of ceramics but also to understand their physical properties.

As stated earlier, there exist three aspects of materials research: property, structure and manufacture. Although it has been pointed out that it is no use considering only one area of materials research, the most important aspect of ceramics research lies in their structure. Structure is the key to relate property to manufacture. Because of the diversity of ceramic structures, their physical properties are also quite diverse and their manufacturing techniques are rich in variety. In this book an attempt has been made to focus on structures and to explain physical properties and manufacturing methods as well.

1.4 FUTURE PERSPECTIVE

Impetus for the development of new materials changes with time. Until the latter half of the 1970s materials were mass produced and their consumption increased as they became cheaper. Materials amenable to mass production were judged to be superior. But mass production consumed a large amount of energy and natural resources and burdened our environment by creating substantial pollution and waste.

A new materials age dawned in the latter half of the 1970s. The impetus for the development of new materials shifted toward finding new materials with superior properties. Stronger materials and those more resistant to harsher environments were increasingly sought. So-called structural ceramics were developed in response to such demands. However, this prevailing philosophy needs to be revised. One of the major problems is in the difficulty of recycling these structural ceramics after use. New materials in the future should be easy to manufacture, safe to use and easy to recycle. Although these demands are self-contradictory it is the responsibility of academia and the technical profession to find such materials. New ceramics must be developed to meet these requirements.

What the authors are advocating here are intelligent materials that have simple compositions and structures, and are capable of self-diagnosis, self-control and self-repair. Functional and structural materials which satisfy these requirements are those unique materials whose functions are inherent in their structure. The authors wish to see the use of these materials and the establishment of such an academic discipline in the future.

2

STRUCTURES OF CERAMICS

In general, in solid-state chemistry the word 'structure' implies a crystal structure. However, in ceramics 'structure' covers a wide range of ceramic structures, ranging from a very minute electron orbit in the order of Ångstroms (10^{-10} m $= 0.1$ nm) to an assembly of grains in the order of micrometers ($1 \sim 10\, \mu$m). As will be discussed in Chapter 4, this wide range arises from the fact that physical properties of ceramics are affected by a great variety of structural aspects. The structures of ceramics can be classified into two main categories, i.e. structures in ceramic grains and structures between ceramic grains and their surroundings. The former includes electron orbits, chemical bonds, crystal structures, lattice defects and dislocations. The latter comprises surfaces, interfaces, grain boundaries, grain sizes, grain size distributions, morphologies, porosity, pore sizes and grain orientations. Since we cannot cover all these subjects in detail, an attempt has been made to describe only the subjects considered to be important from a chemical viewpoint.

2.1 CHEMICAL BONDS IN CERAMICS

Ceramics are defined as non-metallic, inorganic solid materials which have either ionic or covalent bonds. Their characteristics are as follows:

(1) Ionic bonds are formed between quasi-spherical bodies of anions and cations. The bonding is governed by a central force called a Coulombic force which is a function of the distance between anions and cations only, i.e. $\phi = f(r)$ and $\phi \neq f(\theta, \varphi)$. Since, in general, anions are larger than cations, anions form a close-packed structure and cations fill its interstices. The size of cations determines the selection of the interstices. The number of anions which surround a cation is called the coordination number.

(2) Covalent bonds are formed between atoms. These bonds are governed by a non-central force which is a function of the distance and angle between

atoms, i.e. $\phi = f(r, \theta, \varphi)$. The shapes of the covalent bonds are controlled by the contribution of electron orbits on the covalent bonds.

In general, ionic bonds tend to form a close-packed structure and covalent bonds a less packed structure. Materials with ionic bonds tend to have higher thermal coefficients of expansion than those with covalent bonds. Figure 2.1 shows the geometric arrangements of anions and cations with various ionic radii and coordination numbers and Fig. 2.2 the relationships between electron orbits and shapes of covalent bonds. The radii of various ions are summarized in Table 2.1.

Triangular, tetrahedral and octahedral structures are seen in materials with both ionic and covalent bonds. The bonds which form a close geometric arrangement when viewed either covalently or ionically are quite stable. An example of such a structure is a tetrahedron of SiO_4 (a coordination number of four and sp^3 mixed orbits). On the other hand, the bonds whose coordination numbers based on their ionic radii differ from those based on covalent bonding are quite unstable (ZnO has this class of bond). From consideration of ionic radii a coordination number of six is anticipated, but in reality it has a coordination number of four due to the presence of sp^3 mixed orbits.

The coordination numbers of transition metal ions are further complicated due to the stabilizing effect of a given geometric arrangement by the splitting of

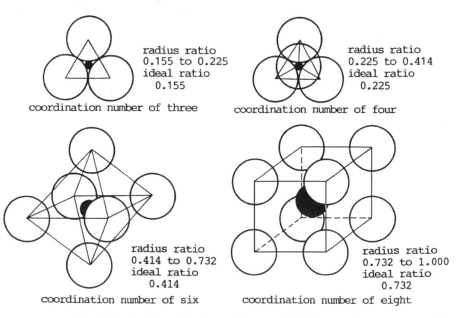

radius ratio
0.155 to 0.225
ideal ratio
0.155

coordination number of three

radius ratio
0.225 to 0.414
ideal ratio
0.225

coordination number of four

radius ratio
0.414 to 0.732
ideal ratio
0.414

coordination number of six

radius ratio
0.732 to 1.000
ideal ratio
0.732

coordination number of eight

Fig. 2.1 Radius ratios for various coordination numbers

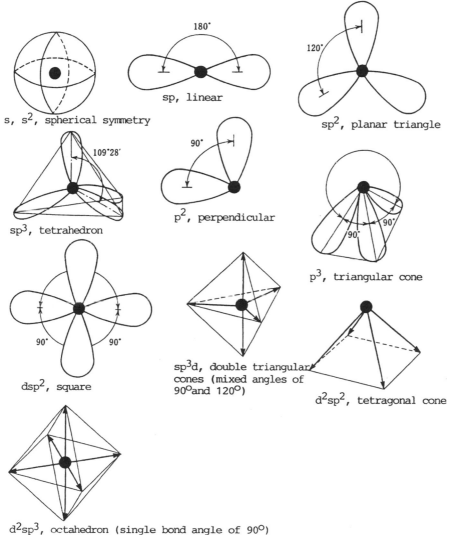

Fig. 2.2 Various types of covalent bonds

electron orbits in the coordination field or crystal field. Cr^{3+} with d^3 electrons and Ni^{2+} with d^6 electrons occupy stable octahedral positions. Co^{2+} with d^7 electrons and Mn^{3+} with d^4 electrons are likely to occupy octahedral positions ionically, but tend to occupy tetrahedral positions from consideration of the crystal field. Due to the fact that electrons in d orbits do not affect Zn^{2+} with a closed shell structure of d^{10}, Zn^{2+} ions occupy tetrahedral positions mainly

Table 2.1   Ionic radii of various cations

| Cation | Coordination number | | | | | Notes |
	3	4	6	8	12	
B^{3+}	0.21	0.22	0.211 0.216	F_3 O_3		
Cl^{7+}		0.25				
S^{6+}		0.29	0.306	F_4^3		
P^{5+}		0.33	0.316	O_4^3		
Be^{2+}		0.33				
Br^{7+}		0.37				
Si^{4+}		0.40	0.47			
Se^{6+}		0.40				
Mn^{7+}		0.44				$Mn^{4+}0.57(4)$
As^{5+}		0.44				$As^{4+}0.49(4)$, $As^{3+}0.55(4)$
Al^{3+}		0.49	0.51			
Cr^{6+}		0.49				$Cr^{4+}0.60(4)$
(As^{4+})		(0.49)				$As^{5+}0.44(4)$
Ge^{4+}		0.50	0.53			
(Cu^{3+})			0.54			$Cu^{2+}0.72(6)$
Tc^{7+}		0.53	0.56			
As^{3+}		0.55	(0.57)	0.563	F_6^4	$As^{4+}0.49(4)$, $As^{5+}0.44(4)$
Fe^{4+}		(0.57)	0.58	0.580	O_6^4	$Fe^{3+}0.64(6)$
V^{5+}		0.56	0.59			$V^{4+}0.66(6)$, $V^{3+}0.74(6)$
Mn^{4+}		(0.57)	0.60			$Mn^{7+}0.44(4)$, $Mn^{3+}0.70(6)$
(Ni^{3+})		(0.57)	(0.60)			$Ni^{2+}0.69(6)$
W^{6+}		0.59	0.62			$W^{4+}0.72 (6)$
Ga^{3+}		0.59	0.62			
Mo^{6+}		0.59	0.62			
Sb^{5+}		0.59	0.62			$Sb^{4+}0.69(6)$
(Cr^{4+})		(0.60)	(0.63)			$Cr^{6+}0.49(4)$, $Cr^{3+}0.69(6)$
Co^{3+}		(0.60)	0.63			$Co^{2+}0.72(6)$
Fe^{3+}		(0.61)	0.64			$Fe^{4+}0.58(6)$, $Fe^{2+}0.74(6)$
Pd^{4+}			0.65			$Pd^{2+}0.80(6)$
Pt^{4+}			0.65			$Pt^{2+}0.80(6)$
Mg^{2+}		(0.62)	0.66			
(V^{4+})			(0.66)			$V^{5+}0.59(6)$, $V^{3+}0.74(6)$
Ru^{4+}			0.67			$Ru^{2+}0.91(6)$, $Ru^{3+}0.80(6)$
Ta^{5+}			0.68			
Ti^{4+}		(0.64)	0.68			$Ti^{3+}0.76(6)$
Li^{+}		(0.64)	0.68			
Rh^{3+}			0.68			$Rh^{2+}0.83(6)$
(Sb^{4+})			0.69			$Sb^{5+}0.62(6)$, $Sb^{3+}0.76(6)$
Nb^{5+}			0.69			$Nb^{4+}0.74(6)$
Ni^{2+}		(0.65)	0.69			$Ni^{3+}0.60(6)$
Cr^{3+}		(0.65)	0.69			$Cr^{4+}0.60(4)$, $0.63(6)$
Mn^{3+}		(0.65)	0.70			$Mn^{4+}0.60(6)$, $Mn^{2+}0.80(6)$

(*continued*)

Table 2.1 (continued)

| Cation | \multicolumn{5}{c}{Coordination number} | Notes |
	3	4	6	8	12	
Np^{7+}			0.71			
Sn^{4+}			0.71			$Sn^{2+}0.93(6)$, $Sn^{3+}0.81(6)$
W^{4+}			0.72			$W^{6+}0.62(6)$
Co^{2+}		(0.68)	0.72			$Co^{3+}0.63(6)$
Cu^{2+}		(0.68)	0.72			$Cu^{3+}0.54(6)$, $Cu^{+}0.96(6)$
Bi^{5+}			0.74			$Bi^{4+}0.85(6)$
V^{3+}			0.74			$V^{4+}0.66(6)$, $V^{5+}0.59(6)$
Fe^{2+}		(0.70)	0.74			$Fe^{3+}0.64(6)$
Zn^{2+}		0.71	0.74			$Zn^{+}0.93(4)$
(Nb^{4+})			(0.74)			$Nb^{5+}0.69(6)$
Sb^{3+}			0.76			$Sb^{4+}0.69(6)$, $Sb^{5+}0.62(6)$
Ti^{3+}			0.76			$Ti^{4+}0.68(6)$
Hf^{4+}			0.78	(0.81)		
Zr^{4+}			0.79	0.82		
Pd^{2+}			0.80			$Pd^{4+}0.65(6)$
Pt^{2+}			0.80			$Pt^{4+}0.65(6)$
Mn^{2+}			0.80			$Mn^{3+}0.70(6)$
U^{6+}			0.80			$U^{5+}0.87(6)$
Ru^{3+}			0.80			$Ru^{2+}0.91(6)$, $Ru^{4+}0.67(6)$
In^{3+}			0.81			$In^{+}1.15(6)$
Tb^{4+}			0.81			$Tb^{3+}0.93(6)$
Sc^{3+}			0.81	(0.84)	(0.87)	
(Sn^{3+})			0.81			$Sn^{4+}0.71(6)$, $Sn^{2+}0.93(6)$
Pu^{5+}			0.83			
Gd^{4+}			0.83	(0.86)		$Gd^{3+}0.97(6)$
Rh^{2+}			0.83			$Rh^{3+}0.68(6)$
Pb^{4+}			0.84	(0.87)		$Pb^{2+}1.19(6)$
(Bi^{4+})			0.85			$Bi^{5+}0.74(6)$, $Bi^{3+}0.96(6)$
Np^{5+}			0.85			$Np^{4+}0.96(6)$
Lu^{3+}			0.85		(0.92)	$Lu^{2+}0.89(6)$
Eu^{4+}			0.85	(0.88)		$Eu^{3+}0.98(6)$
Sm^{4+}			0.86	(0.89)		$Sm^{3+}0.98(6)$
Yb^{3+}			0.86	(0.89)	(0.93)	
U^{5+}			0.87			$U^{6+}0.80(6)$, $U^{4+}0.97(6)$
Tm^{3+}			0.87	(0.90)	(0.94)	
Pm^{4+}			0.88	(0.91)	(0.95)	$Pm^{3+}1.02(6)$
Am^{4+}			(0.89)	0.92		$Am^{3+}1.03(6)$
Er^{3+}			0.89	(0.92)	(0.96)	
Pa^{5+}			0.89			
Lu^{2+}			(0.90)	0.93	(0.97)	$Lu^{3+}0.85(6)$
Nd^{4+}			0.90	(0.94)	(0.97)	$Nd^{3+}1.04(6)$
Ho^{3+}			0.91	(0.95)	(0.98)	
Ru^{2+}			0.91			$Ru^{4+}0.67(6)$

(continued)

Table 2.1 (*continued*)

Cation	3	4	6	8	12	Notes
Pu^{4+}			(0.92)	0.96		$Pu^{3+}1.12(8)$
Pr^{4+}			0.92	(0.96)		
Y^{3+}			0.92	(0.96)	(0.99)	
Dy^{3+}			0.92	(0.96)	(0.99)	$Dy^{2+}0.99(6)$
Sn^{2+}			0.93	(0.97)		$Sn^{4+}0.71(6), Sn^{3+}0.81(6)$
Tb^{3+}			0.93	(0.97)	(1.00)	$Tb^{4+}0.81(6), Tb^{2+}1.00(6)$
Ce^{4+}			0.94	(0.98)	(1.01)	$Ce^{3+}1.07(6)$
Tl^{3+}			0.95			
Np^{4+}			0.95			$Np^{3+}1.14(8), Np^{5+}0.85(6)$
Cu^{+}			0.96	(1.00)		$Cu^{2+}0.72(6)$
Bi^{3+}			0.96		(1.03)	$Bi^{4+}0.85(6), Bi^{5+}0.74(6)$
Gd^{3+}			0.97	(1.01)	(1.04)	$Gd^{2+}1.07(6), Gd^{4+}0.83(6)$
Na^{+}			0.97	(1.01)	(1.04)	
Cd^{2+}			0.97	1.01	(1.04)	
U^{4+}			(0.97)	1.01	(1.04)	$U^{5+}0.87(6), U^{3+}1.16(8)$
Eu^{3+}			0.98	(1.02)	(1.06)	$Eu^{4+}0.85(6), Eu^{2+}1.08(6)$
Pa^{4+}			0.98			$Pa^{3+}1.17(8)$
Sm^{3+}			0.98	(1.02)	(1.06)	$Sm^{4+}0.86(6), Sm^{2+}1.10(6)$
Dy^{2+}			0.99	1.03	1.07	$Dy^{3+}0.92(6)$
Zn^{+}		(0.93)	(0.99)		(1.07)	$Zn^{2+}0.71(4)$
Ca^{2+}		0.995	(0.99)	1.03	1.07	
Tb^{2+}	F_8^6		(1.00)	1.04		$Tb^{3+}0.93(6)$
Pm^{3+}			1.02	(1.06)		$Pm^{4+}0.88(6)$
Th^{4+}	O_8^6	1.026	(1.02)	1.06	(1.10)	$Th^{3+}1.19(8)$
Am^{3+}			(1.03)	1.07		$Am^{4+}0.89(6)$
Nd^{3+}			1.04	(1.08)	(1.12)	$Nd^{4+}0.90(6)$
Pr^{3+}			1.06	(1.10)	(1.14)	$Pr^{4+}1.19(8)$
Ac^{4+}			(1.06)	1.10		$Ac^{3+}1.22(8)$
Gd^{2+}			(1.07)	1.11	(1.15)	$Gd^{3+}0.97(6)$
Ce^{3+}			1.07	1.11	1.15	$Ce^{4+}0.94(6), Ce^{2+}1.22(8)$
Eu^{2+}			(1.08)	1.12	(1.17)	$Eu^{3+}0.98(6)$
Pu^{3+}			(1.08)	1.12		$Pu^{4+}0.96(8)$
Sm^{2+}			(1.10)	1.14		$Sm^{3+}0.98(6)$
Np^{3+}			(1.10)	1.14		$Np^{4+}0.95(6)$
Hg^{2+}			1.10	1.14	(1.19)	
U^{3+}			(1.12)	1.16		$U^{4+}1.01(8)$
Pm^{2+}			(1.12)	1.16		$Pm^{3+}1.02(6)$
Pa^{3+}			(1.13)	1.17		$Pa^{4+}0.98(6)$
La^{3+}			1.14	1.18	1.23	$La^{2+}1.28(8)$
Pr^{2+}			(1.14)	1.19		$Pr^{3+}1.06(6)$
Th^{3+}			(1.14)	1.19		$Th^{4+}1.06(8)$
In^{+}			1.15	1.20	1.24	$In^{3+}0.81(6)$
Sr^{2+}			1.16	(1.21)	1.25	

(*continued*)

Table 2.1 (*continued*)

Cation			Coordination number			Notes
	3	4	6	8	12	
Ce^{2+}			(1.17)	1.22		$Ce^{3+}1.07(6)$
Ac^{3+}			(1.17)	1.22		$Ac^{4+}1.10(8)$
Pb^{2+}			(1.19)	(1.24)	(1.29)	$Pb^{4+}0.84(6)$
La^{2+}			(1.23)	1.28	_ _ _ _	$La^{3+}1.18(8)$
Ag^+			(1.26)	1.31		
K^+			1.33	(1.38)	1.44	
Ba^{2+}	F^8_{12}	1.36	1.36	1.43	1.47	
Tl^+	O^8_{12}	1.40	(1.48)	(1.54)	1.60	
Rb^+			1.52	(1.56)	1.60	
Cs^+			1.70	(1.77)	1.82	

Notes: O^n_m in the coordination number column indicates that cations with ionic radius smaller than that indicated have n coordination number.
Cations with ionic radius larger than that indicated have m coordination of oxygen ions. F^n_m indicates similarly for F^- ions.

because of the sp^3 mixed orbits. Mn^{2+} and Fe^{3+} have half-filled d orbits (d^5) and occupy either tetrahedral or octahedral positions. Ionically, these cations should occupy octahedral positions. Whether transition metal ions occupy either tetrahedral or octahedral positions depends on the presence of other ions and the structure of a host crystal. Based solely on the crystal field, they tend to preferentially occupy octahedral positions as indicated below:

$$\leftarrow \text{Octahedral} \qquad\qquad\qquad \text{Tetrahedral} \rightarrow$$

	Cr^{3+}	Mn^{3+}	Ni^{2+}	Cu^{2+}	V^{3+}	Ti^{3+}	$\sim Co^{2+}$	Fe^{2+}	$Mn^{2+}, \sim Fe^{3+}$	Zn^{2+}
No. of d electrons	(3)	(4)	(8)	(9)	(2)	(3)	(7)	(6)	(5)	(10)

Cations in the most stable crystal structures have their ionic radii which are very close to the ideal radii for given coordination numbers. The larger the deviation from the ideal radii, the less stable is the crystal structure (see Table 2.2). It is easier to dissolve foreign atoms or ions, which tend to lessen the deviation, in a crystal, but it is difficult to do so when they tend to increase the deviation.

2.2 PACKING OF SOLID SPHERES

There are a number of ways that identical atoms or anions in an ionic crystal can be arranged periodically. The simplest arrangement is a simple cubic packing. This packing occurs when every corner of a cube is occupied by an

Table 2.2 Coordination numbers of cations in various oxide crystals

Oxide	Cation radius (Å)	Coordination number anticipated from the ionic radius	Actual coordination number	Deviation of ionic radius from the ideal radius	Notes
BeO	0.33 (4)	4	4	+0.014	Hint of sp^2
MgO	0.66 (6)	6	6	+0.08	Tends to have tetrahedral coordination in spinel. Has the highest melting point for divalent cation oxides
CaO	0.99 (6)	6 (8)	6 (8)	+0.41	Tends to form peroxide, CaO_2
SrO	1.16 (6)	8	6	+0.58	Tends to form peroxide, SrO_2
BaO	1.43 (6)	12	6	+0.85	Tends to form peroxide, BaO_2
NiO	0.69 (6)	6	6	+0.11	d sp mixed orbitals and octahedrally coordinated
CoO	0.72 (6)	6	6	+0.14	d^2 sp^3 mixed orbitals, octahedrally coordinated (tetrahedrally coordinated in spinels $d^7 = e^4t^3$) ($d^8 = t^6e^2$)
ZnO	0.74 (6)	6	4	+0.424	sp^3 mixed orbitals
B_2O_3	0.21 (3)	3	3 (4)	−0.006	Tetrahedral coordination with Na_2O
Al_2O_3	0.51 (6) / 0.49 (4)	4 (6)	6 (4)	−0.07 (6) / +0.174 (4)	Octahedrally coordinated in α-alumina, but has unequal Al–O distances. Coordinated octahedrally and tetrahedrally in transition type polymorphs such as γ-alumina
Bi_2O_3	0.96 (6) / 1.00 (8)	6	(8) 8~2	−0.026 (8) / +0.38 (6)	Has CaF_2 (8) structure
SiO_2	0.40 (4)	4	4	0.084	sp^3 mixed orbital and tetrahedral coordination
TiO_2	0.68 (6)	6	6	0.10	
ZrO_2	0.79 (6) / 0.82 (8)	6	8	+0.21 (6) / −0.206 (8)	Has unequal Zr–O distances in ZrO_2; CaO and Y_2O_3 are added to stabilize the cuboidal coordination
ThO_2	1.06 (8)	8	8	+0.034	Has the highest melting point in oxides

atom or an ion (Fig. 2.3). A large number of the cubes are arranged three-dimensionally. Po is an example which has a simple cubic structure. CsCl is an ionic example whose chlorine ions form a simple cubic structure with cesium ions at the center of the cubes (Fig. 2.4).

Consider a two-dimensional arrangement of solid spheres (Fig. 2.5(a)). The center of the spheres will be called A and two adjacent interstices B and C. When a second layer of the solid spheres is laid over the first, the solid spheres occupy either the B or C positions. Let the occupied positions be B and the unoccupied positions C. When a third layer of the solid spheres is laid over the second layer, the solid spheres occupy either A or C positions. When the layers are stacked in a sequence ABABAB . . ., the structure is called hexagonal close packing (hcp). On the other hand, when the layers are stacked in a sequence

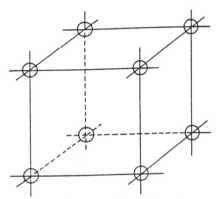

Fig. 2.3 Simple cubic packing (Po metal)

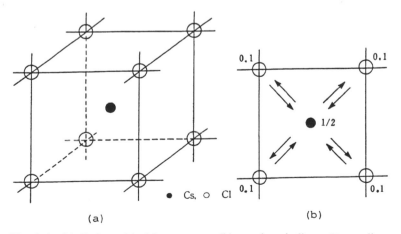

(a)   (b)

Fig. 2.4 (a) Cesium chloride structure; (b) numbers indicate Z coordinates

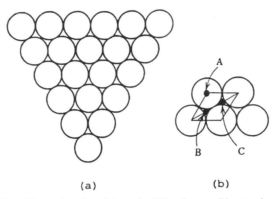

(a)   (b)

Fig. 2.5 (a) Two-dimensional packing of solid spheres; (b) atomic positions for second and third layers

ABCABC . . ., the structure is called either cubic close packing (ccp) or face centered cubic packing (fcc).

When a structure with identical ions at every corner and also at the center of every face of the cubes is viewed along a diagonal direction of the cube, the structure exhibits a stacking sequence ABCABC . . . . Hence, ccp is identical to fcc. Cartesian coordinates of a face centered cubic lattice are shown in Fig. 2.6(b).

Cubic close packed structures are found in Al, Ag, Au, and Cu, which are good ductile metals, and Ar and Ne, which are noble gas crystals. Hexagonal

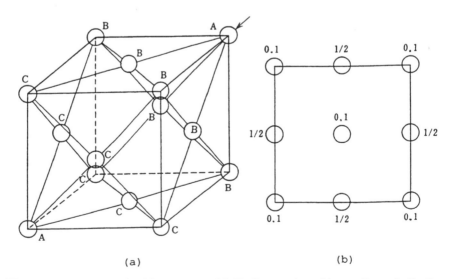

(a)   (b)

Fig. 2.6 Face centered cubic structure. (a) Bird's-eye view; (b) coordinate indication

close packed structures are found in Ni and Zn. These metals are somewhat less ductile than the ccp metals. Ionic compounds which crystallize in either ccp or hcp structures will be discussed in detail later.

2.3 CLASSIFICATION OF CRYSTAL STRUCTURES FROM THE COORDINATION NUMBERS OR THE SEQUENCES OF ION PACKING

2.3.1 Compounds with A:X = 1:1

2.3.1.1 THE COORDINATION NUMBER OF THREE

h-BN belongs to this structural group. As indicated in Fig. 2.7, this structure consists of a two-dimensional network of hexagonal rings of BN due to the fact that BN forms a completely filled valence shell. Because the weak van der Waals force acts between the layers, the layers slide quite easily.

Graphite is another example of this structural group. In addition to the low friction characteristic, graphite is electrically conductive due to the presence of π electrons and is black in color. On the other hand, h-BN is white and is called either white carbon or white graphite. Because of the reaction resistance of h-BN with Fe and its white color, h-BN is often used as a lubricant in place of graphite. The properties of both h-BN and graphite are summarized in Table 2.3. Lubricity is the commonalty of these two materials.

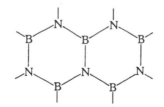

Fig. 2.7 Layer structure of boron nitride

Table 2.3 Properties of h-BN and graphite

Structure with coordination number of three	h-BN	Graphite
Electrical property	Insulating	semiconducting
Color	White	Black
Reactivity with Fe	Small	Large

2.3.1.2 THE COORDINATION NUMBER OF FOUR

(1) A structure which has anions in ccp and cations in a half of the resulting tetrahedral interstices is called a zinc blende (β-ZnS) structure (Fig. 2.8). When h-BN is heated above 1000°C under pressure ($\sim 10^4$ atm), the structure of h-BN converts to a tetrahedral coordination. AgI, β-ZnS and β-SiC all have this structure.

Diamond, Si and Ge have a similar structure in which identical atomic species occupy ccp positions as well as the interstices. Because of the absence of π electrons, diamond is an electrical insulator. The relationship between diamond and graphite is identical to that between the high-pressure phase of BN (c-BN) and h-BN. Like diamond, c-BN has been employed to make cutting and grinding tools, but because of the poor reactivity of BN with Fe, BN tools can be used to machine a wide variety of Fe-bearing materials.

(2) A structure which has anions in hcp and cations in half of the resulting interstices is call the wurtzite (α-ZnS) structure (Fig. 2.9). This can be

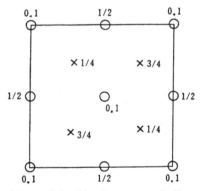

Fig. 2.8 Atomic positions of zinc blende structure indicated by coordinates. o Positions of A atoms; X positions of B atoms

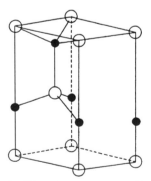

Fig. 2.9 Wurtzite structure. ● Cation; o anion

found in α-SiC, BeO, α-ZnS, ZnO and AlN. BN forms a wurtzite as well as a zinc blende structure.

In ZnO the ionic radius of Zn^{2+} is large enough to be coordinated with six O^{2-}. But because of the presence of sp^3 mixed orbitals in Zn–O bonds, Zn ions occupy tetrahedral interstices. In this structure it is difficult to incorporate cations with larger ionic radii, which tend to increase the deviation from the ideal anion-to-cation ratio. On the other hand, it is easier to incorporate cations with smaller ionic radii.

2.3.1.3 THE COORDINATION NUMBER OF SIX

(1) The structure with anions in ccp and cations at all resulting octahedral interstices occurs for NaCl, and is commonly known as the rock salt structure (Fig. 2.10). Crystals having this structure are LiF, NaH, KCl, AgCl, MgO, CaO, SrO, BaO, PbS, TiC, MnO, FeO, CoO, and NiO. The factors which determine the stability of octahedral coordination in transition metal oxides such as MnO, FeO, CoO, and NiO are the ionic radius and the presence of subcoordination field or crystal field due to the presence of d electrons. MgO is the only alkaline earth oxide which has octahedral coordination. Since other alkaline earth cations are too large for an octahedral coordination, they tend to form peroxides such as BaO_2 and SrO_2. In these compounds oxide ions (O^{2-}) are replaced with peroxide ions (O_2^{2-}) that have much larger ionic radii. Because of the non-spherical symmetry of O_2^{2-} ions, these peroxides tend to form a deformed structure which is tetragonal and this is commonly known as the CaC_2 structure.

(2) The structure with anions in hcp and cations in all resulting octahedral interstices occurs in NiAs, and is commonly known as the nickel arsenide structure. Other crystals having this structure are FeS and CrS.

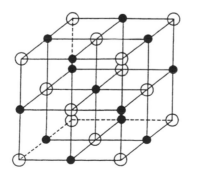

Fig. 2.10 Rock salt structure. ● Cation; ○ anion

2.3.1.4   THE COORDINATION NUMBER OF EIGHT

The structure with anions at every corner of cubes and cations at the center of the cubes (body center) occurs in CsCl, and is commonly known as the cesium chloride structure. Crystals having this structure are CsBr, CsI, TiCl, and NH_4Cl.

2.3.2   Compounds with A:X = 1:2

2.3.2.1   THE COORDINATION NUMBER OF FOUR

Cations, A, occur in the interstices at the centers of octahedra formed by anions, X. Since the octahedra are bonded together by sharing only the corners, there exists a substantial amount of variation in the A—X—A angle. Because of the variation in the bonding angle, it is possible to see a large variation in the structure. (Structures with identical chemical composition but different crystal structures are commonly called polymorphs.) SiO_2 is a good example of this crystal group and crystallizes in quartz, tridymite, cristobalite and glasses. In cristobalite, Si ions occupy the carbon positions of diamond. Oxide ions are present in the middle of the Si–Si bonds. Tridymite is similar to the wurtzite structure. BeF_2 has polymorphs similar to SiO_2, but the temperatures at which they change structures (transformation temperatures) are lower than those of SiO_2. Thus BeF_2 has been investigated as an alternative to SiO_2 for studying phase transformation of SiO_2. Ice (H_2O) has a reverse structure of SiO_2.

2.3.2.2   THE COORDINATION NUMBER OF SIX

The structure with anions in hcp and cations in one half of the resulting octahedral interstices occurs in CdI_2, and is commonly known as the CdI_2 structure, which is a layer structure as indicated in Fig. 2.11. The layers without cations are bonded with the van der Waals force and thus slide quite easily.

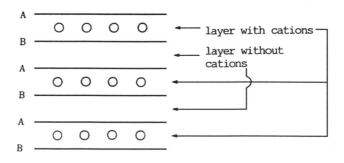

Fig. 2.11 CdI_2 structure

Another example of this structure is $Mg(OH)_2$, commonly called brucite. In $Al(OH)_3$, two thirds of the octahedral interstices in the CdI_2 are occupied by Al^{3+}. When OH^- layers in $Al(OH)_3$ are arranged ABABAB . . ., identically to CdI_2, the structure is called bayerite. In addition, gibbsite has an arrangement of {AB}{BA}{AB}{BA} . . . and nordstrandite an arrangement of {ABAB} {BABA}{ABAB}{BABA} . . . .

The structure with cations at every corner of a cube and also at the center of the cube and with anions to surround these cations octahedrally occurs in both rutile (TiO_2) and SnO_2, and is commonly known as either the rutile or the SnO_2 structure (Fig. 2.12). Anions are packed in a distorted hcp structure. The compounds with this structure have a variety of interesting characteristics. TiO_2 has been used for corrosion-resistant materials, coating materials (for gas sensors, based on its non-stoichiometric characteristics and IR reflection for glasses), pigment materials, electrode materials, and dielectric materials. SnO_2 has been used for translucent semiconducting materials and coating materials of IR reflection for glasses and for gas sensors. The mechanism of the SnO_2 sensors is based on the absorption and desorption of gases on SnO_2. Oxides having this structure are VO_2, β-MnO_2, RuO_2, OsO_2, IrO_2, GeO_2, PbO_2, NbO_2, and TaO_2. In these oxides there is a series of oxides, designated by M_nO_{2n-1}, with a sheared structure. Basically this structure results when O^{2-} is removed by sharing a network of TiO_6 octahedra. Examples of sheared structures are Ti_3O_5, Ti_4O_7 and Ti_5O_9. In the above notation TiO_2 can be viewed as a sheared structure with $n = \infty$.

2.3.2.3 THE COORDINATION NUMBER OF EIGHT

The structure resulting from the removal of every other cation in the CsCl structure shown in Fig. 2.13 occurs in CaF_2, and is commonly known as the

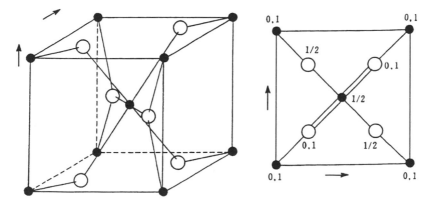

Fig. 2.12 Rutile structure. ● A; ○ X

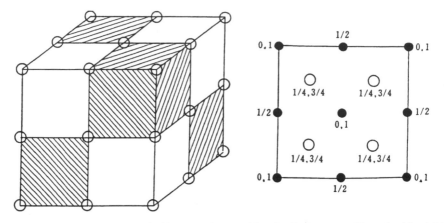

Fig. 2.13 CaF$_2$ structure (o indicates anion positions). Cations coordinated with eight anions are located at the center of the hatched cubes but the other cubes do not contain cations

fluorite structure. In this structure cations occupy every corner of a cube as well as every face center. Eight anions are placed in the face centered cubic lattice.

Compounds having this structure are SrF$_2$, ThO$_2$, and ZrO$_2$. A major characteristic of the compounds with the fluorite structure is the high mobility of anions. ThO$_2$ has the highest melting point in oxides. The ionic radius of Zr^{4+} in ZrO$_2$ is rather smaller than that of an 8 fold coordination and thus the structure of ZrO$_2$ is distorted. In order to stabilize the structure of ZrO$_2$, it is necessary to incorporate CaO and/or Y$_2$O$_3$. The solid solution of these oxides causes the creation of oxide ion vacancies as indicated by Ca$_x$Zr$_{1-x}$O$_{2-x}$, which makes zirconia a good oxide ion conductor.

Li$_2$O, Na$_2$O, K$_2$O and Rb$_2$O have the inverse CaF$_2$ structure. One of the polymorphs of Bi$_2$O$_3$, δ-Bi$_2$O$_3$, has a structure resulting from the random removal of one fourth of anions in the CaF$_2$ structure.

2.3.3 Compounds with A:X = 2:3

2.3.3.1 THE COORDINATION NUMBER OF SIX

The structure with anions in hcp and cations in two thirds of the resulting octahedral interstices occurs in corundum (α-Al$_2$O$_3$), and is commonly known as the corundum structure, shown schematically in Fig. 2.14. Oxides having this structure are α-Fe$_2$O$_3$, Cr$_2$O$_3$, Ti$_2$O$_3$ and V$_2$O$_3$. Oxide ion octahedra are bonded together by sharing their faces. This is shown in Fig. 2.15. It is rare to find crystals which share faces of the polyhedra.

According to Pauling,

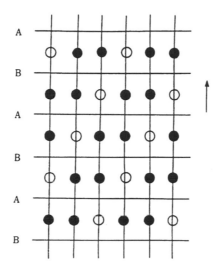

Fig. 2.14 Corundum structure. ● positions occupied by cations; ○ positions not occupied by cations

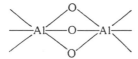

Fig. 2.15 Schematic of Al–O octahedral covalent bonds in α-Al_2O_3

(1) The coordination number of cations is determined by the ratio of ionic radii between cations and anions in ionic crystals.
(2) It is necessary to maintain electrical neutrality in the crystals.
(3) Crystals which share corners of the coordination polyhedra are the most stable and those that share edges and faces of the coordination polyhedra are progressively less stable.

According to rule (3) above, the corundum structure is assumed to be unstable. Thus Al_2O_3 and Fe_2O_3 tend to crystallize into another structure which can be exemplified by γ-Al_2O_3 and γ-Fe_2O_3. These oxides have crystal structures similar to spinel structures which will be discussed later. These oxides have coordination numbers of both four and six.

Corundum is the name of the mineral α-Al_2O_3. Sapphire is a solid solution between α-Al_2O_3 and either Ti_2O_3 or Fe_2O_3 and has a blue to purple color. Ruby is a solid solution between α-Al_2O_3 and Cr_2O_3 and has a red color. The word 'sapphire' is also used to indicate high-purity single crystals of α-Al_2O_3

without any coloration (for example, SOS (silicon on sapphire): a silicon device built on a high-purity single-crystal substrate).

Pure alumina is hard, resistant to corrosion, electrically insulating, thermally conducting, resistant to heat and translucent. Because of these unique characteristics, alumina has been used as IC substrates, discharge tubes in high-pressure sodium lamps, bearings, crucibles, cutting tools, grinding, and polishing materials and jewelry. γ-Al_2O_3 is porous and has been used as catalyst carriers and absorbents.

Magnetically, α-Fe_2O_3 is very much different from γ-Fe_2O_3. The former is strongly antiferromagnetic and does not magnetize when a magnetic field is applied. On the other hand, the latter is ferrimagnetic and magnetizes when a magnetic field is applied. The magnetization remains even when the magnetic field is removed. Because of this unique property, needle-shaped γ-Fe_2O_3 has been used in magnetic tapes for recording.

In $MgTiO_3$ Mg and Ti occupy every other Al position in the corundum structure and this is commonly known as the ilmenite structure. $FeTiO_3$ also has the ilmenite structure.

2.3.4 Compounds with a Chemical Formula of AB_2X_4

An example of the crystals in this specific class is the spinel structure, which is cubic and consists of eight smaller cubes. As shown in Fig. 2.16(a), the smaller cubes can be classified into two types, I and II. In subcubes I, the first cations occupy half of the eight corners diagonally, as shown by the solid circles in Fig. 2.17(a), as well as at the center of the subcube. The cations at the center of the subcube are surrounded by nuclei of four anions which are located diagonally. Thus the cations are tetrahedrally coordinated with four anions.

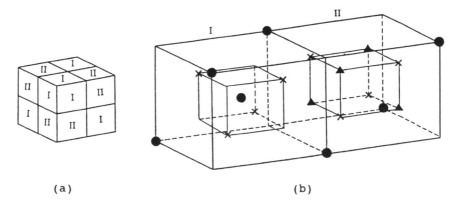

(a) (b)

Fig. 2.16 Schematic of the spinel structure. (a) Unit lattices; (b) partial lattices of the spinel structure. ● A positions, 8a, tetrahedral; ▲ B positions, 16d octahedral; X oxygen positions, 32e

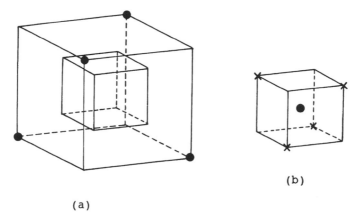

(a)

(b)

Fig. 2.17   (a) Cell I of the spinel structure; (b) a nucleus in the cell

In subcubes II, the first cations, A, occupy half of the eight corners of the subcube diagonally as indicated in Fig. 2.18(a). The subcube contains a nucleus which consists of four second cations located at half of the eight corners diagonally and four anions located at the remaining four corners. Thus the second cations are octahedrally coordinated with anions.

As indicated above, there are eight tetrahedrally coordinated cations, 16 octahedrally coordinated cations and 32 anions in a unit cell of the spinel structure. Thus the unit cell of the spinel structure can be denoted by $(\ )_8[\ ]_{16}X_{32}$, where ( ) indicates a tetrahedrally coordinated cation site and [ ] an octahedrally coordinated cation site. A chemical formula, AB_2X_4, can be indicated more generally by an abbreviated notation, $(\ )[\ ]_2X_4$. The spinel structures which have A cations in the tetrahedral coordination sites and B

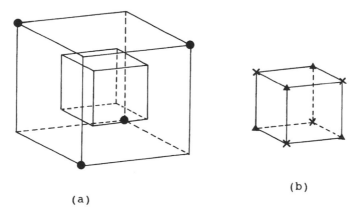

(a)

(b)

Fig. 2.18   (a) Cell II of the spinel structure; (b) a nucleus in the cell

cations in the octahedral coordination sites, and thus can be represented by
$(A)[B]_2X_4$, are called normal spinel structures. On the other hand, the spinel
structures which have B cations in the tetrahedral coordination site and A and
B cations in the octahedral coordination sites can be represented by
$(B)[A,B]X_4$. Thus these are called inverse spinel structures. The occupancy of
A and B cations in either tetrahedral or octahedral coordination sites depends
on the rules discussed in Section 2.2 above. Spinels having the normal structure
are $ZnFe_2O_4$ and $CdFe_2O_4$ and those with the inverse structure are $NiFe_2O_4$
and $MgFe_2O_4$.

In the spinal structure one tetrahedrally coordinated cation, ( ), and two
octahedrally coordinated cations, [ ], are bonded together through an anion as
shown in Fig. 2.19. When cations in both tetrahedral and octahedral
coordination sites have unfilled d electrons and the anions are oxygen ions,
magnetic moments tend to align antiparallel due to the super-exchange
interaction. Since the interaction is strongest at an alignment angle of $180°$, the
magnetic moments in the tetrahedral and octahedral coordination sites tend to
align antiparallel. Thus two cations in the octahedral coordination sites are
aligned parallel. As a result, the spinels tend to have a remnant magnetic
moment, which results from the subtraction of the magnetic moment of the
tetrahedrally coordinated cations from the magnetic moments of the two
octahedrally coordinated cations. The presence of the remnant magnetic
moment is termed ferrimagnetic. The strength of ferrimagnetism depends on
the occupancy of cations in both tetrahedral and octahedral coordination sites.

Solid solutions between $ZnFe_2O_4$, a normal spinel, and various other spinels,
MFe_2O_4, where M denotes cations with various magnetic moment, are shown
in Fig. 2.20. Various cations with unfilled d electrons are listed in Table 2.4.
γ-Fe_2O_3, discussed previously, can be expressed as $(Fe)[Fe_{5/3}\{\ \}_{1/3}]O_4$, where { }
denotes cation vacancies.

Iron oxides with magnetization are classified as ferrites. Since ferrites with
the spinel structure are cubic, the magnetization energy does not depend on
crystal orientation. Hence, these ferrites can be easily magnetized by the

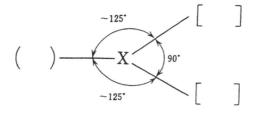

Fig. 2.19 Relationship of cations which exhibit the super-exchange interaction in the
spinel structure

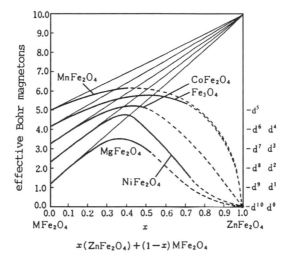

$$x(ZnFe_2O_4) + (1-x)MFe_2O_4$$

Fig. 2.20   Various ferrite solid solutions and their effective Bohr magnetons

Table 2.4   Effective Bohr magnetons in transition metal cations

Cation	Orbital	Ground state	$p(1)$	$p(2)$	Measured	With super-exchange interaction
Ti^{3+}, V^{4+}	$3\,d^1$	$^2D_{3/2}$	1.55	1.73	1.8	1.0
V^{3+}, Cr^{4+}	$3\,d^2$	$^3F_2$	1.63	2.83	2.8	2.0
Cr^{3+}, V^{2+}	$3\,d^3$	$^4F_{3/2}$	0.77	3.87	3.8	3.0
Mn^{3+}, Cr^{2+}	$3\,d^4$	$^5D_0$	0	4.90	4.9	4.0
Fe^{3+}, Mn^{2+}	$3\,d^5$	$^6S_{5/2}$	5.92	5.92	5.9	5.0
Fe^{3+}	$3\,d^6$	$^5D_4$	6.70	4.90	5.4	4.0
Co^{2+}	$3\,d^7$	$^4F_{9/2}$	6.63	3.87	4.8	3.0
Ni^{2+}	$3\,d^8$	$^3F_4$	5.59	2.83	3.2	2.0
Cu^{2+}	$3\,d^9$	$^2D_{5/2}$	3.55	1.73	1.9	1.0

$p(1) = g\{J(J+1)\}^{1/2}$, $p(2) = 2\{S(S+1)\}^{1/2}$ calculated $p(1)$.

application of a small magnetic field. It is also easy to reverse the direction of magnetization. This characteristic is called soft magnetism and has been used in the fabrication of microelectronic elements as well as memory chips and magnetic cores of high-frequency transformers. Ferrites are used in transformers because of their transducer functions. For this particular application, ferrites should have a very small magnetic coercive field and a high apparent magnetic transference number.

There are a group of ferrites which have a chemical composition of $MO \cdot 6Fe_2O_3$, where M are Pb, Ba, and Sr, and have the magnetoplumbite structure. As shown in Fig. 2.21, the magnetoplumbite structure has thin layers of M and O. These oxides are sandwiched between thick layers of oxides which have a crystal structure very similar to the spinel structure. Because of this layer structure, magnetization energies have a wide variation between the parallel and perpendicular directions of the structure. It is necessary to apply a large magnetic field for magnetization, but once magnetized, it is very difficult to erase the magnetization. This class of oxides is useful for the manufacture of magnets.

γ-Fe_2O_3 is used for the manufacturing of magnetic tapes. In order to provide a magnetic force to soft magnetic γ-Fe_2O_3, needle-shaped fine powder has been used for the manufacture of magnetic tapes.

2.3.5   Compounds with a Chemical Formula of ABX₃

The structures with a chemical composition of ABX_3 occur (1) in ilmenite, in which both A and B cations are octahedrally coordinated with X anions, and

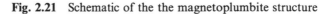

Fig. 2.21   Schematic of the the magnetoplumbite structure

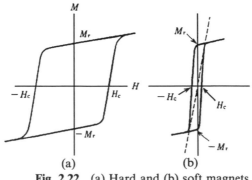

Fig. 2.22   (a) Hard and (b) soft magnets

(2) in perovskite, in which A cations are coordinated with 12 X anions and B cations are octahedrally coordinated with X anions. The former structure is a derivative of the corundum structure, A_2X_3, whose A cation positions are occupied by A and B cations alternatively. Oxides having this structure are $FeTiO_3$, $MgTiO_3$, $MnTiO_3$, and $CoTiO_3$. The latter structure can be illustrated in two ways schematically, which are shown in Fig. 2.23. In the first, a B cation is placed at the center of a cube, which is surrounded with six X anions at the face centers and eight A cations at all corners. In this way it is clear that the A and X ions form a face centered cubic packing. In the second, an A cation is placed at the center of a cube, which makes it clear that the A cation is coordinated with 12 X anions.

In an ideal perovskite crystal ionic radii of A, B, and X ions should satisfy the following relationship:

$$r_A + r_X = (2)^{1/2}(r_B + r_X)$$

where r_A, r_B, and r_X are the ionic radii of A, B, and X ions, respectively. The equation, however, has to be modified to account for the deviation of ionic packing in real crystals from that of the ideal crystal and is given by

$$r_A + r_X = t(2)^{1/2}(r_B + r_X)$$

where t is a tolerance factor, between 0.7 and 1 in most perovskite crystals. Oxides having this structure are $BaTiO_3$, $CaTiO_3$, $SrTiO_3$, $PbTiO_3$, $SrSnO_3$, $SrZrO_3$, and $PbZrO_3$. In addition, there is a series of solid solutions whose chemical compositions are given by

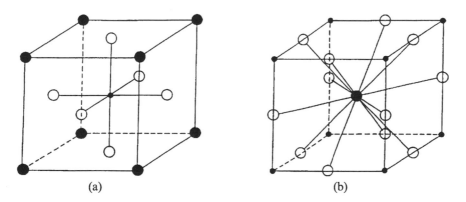

(a) (b)

Fig. 2.23 Perovskite structure. (a) A B cation at the body center (B coordinated with 6 X); (b) an A cation at the body center (A coordinated with 12 X). ● cation: ⊙ B cation; ○ X cation

$$(Ba_xCa_ySr_zPb_{1-x-y-z})\,(Ti_pZr_qSn_rHf_{1-p-q-r})O_3$$

It is further possible to form perovskite crystals by substituting a combination of Na^+ and Bi^{3+} for A sites and a combination of Mg^{2+} and W^{6+} for B sites. There is a large number of perovskite oxides due to the formation of solid solutions among these oxides. The solid solutions are extremely valuable for controlling physical properties and have a significant impact on the commercial applications of the perovskite oxides.

Ideally the perovskite structure is cubic, but it is possible to have some distorted structures. For example, $BaTiO_3$ is cubic at temperatures above 120°C, tetragonal ($a<c$) between 5°C and 120°C, orthorhombic between −80°C and 5°C, and rhombohedral at less than −80°C. These structures are shown in Fig. 2.24.

The tetragonal phase of a perovskite is important for commercial applications. When the $Ti^{4+}-O^{2-}-Ti^{4+}$ bonds in $BaTiO_3$ are viewed along the c-axis, O^{2-} is located off the line connecting the two Ti^{4+}. Thus the center of gravity for anions does not match that for cations. As a consequence, the tetragonal phase of $BaTiO_3$ exhibits a spontaneous polarization. The magnitude of polarization depends on temperature (pyroelectric) and

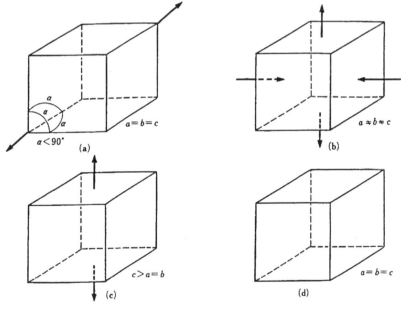

(a)→(b) −80℃. (b)→(c) 5℃. (c)→(d) 120℃

Fig. 2.24 Distortions of various phases of $BaTiO_3$ from the cubic perovskite structure. (a) Rhombohedral; (b) orthorhombic; (c) tetragonal; (d) cubic

mechanical stress (piezoelectric). As indicated in Fig. 2.25, it is possible to flip-flop O^{2-} between two positions by reversing electric potential. This charac-teristic is known as the ferroelectric. As shown in Fig. 2.26, it is possible to change the transition temperature between the cubic and the tetragonal phases by forming solid solutions.

Semiconducting heating elements of the $BaTiO_3$ systems are one example where the wide variation of physical properties by the cubic to tetragonal phase transition can be exploited. As stated previously, it is possible to change the

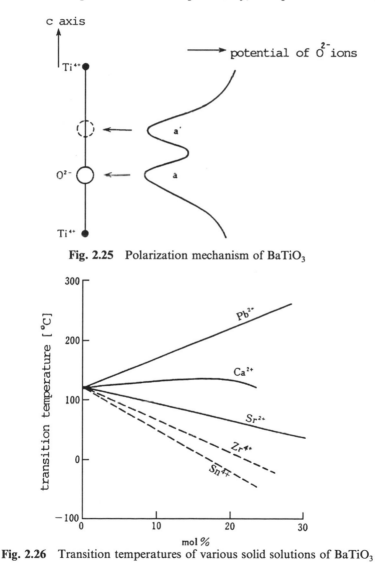

Fig. 2.25 Polarization mechanism of $BaTiO_3$

Fig. 2.26 Transition temperatures of various solid solutions of $BaTiO_3$

transition temperature at which electrical resistivity varies significantly (Fig. 2.27) by the solid solutions.

2.3.6 Structures of Silicates

Glasses, cements and porcelains are important ceramics in modern industries as well as in the traditional arts. The major components of glasses are SiO_2, Na_2O and CaO, of cements are CaO, SiO_2, Al_2O_3 and Fe_2O_3 and of porcelains are SiO_2, Al_2O_3, MgO, K_2O and H_2O. These are the top ten elements in abundance on earth and their Clark numbers are O 49.5%, Si 25.9%, Al 7.56%, Fe 4.7%, Ca 3.39%, Na 3.39%, K 2.40%, Mg 1.93% and H 0.89%. Most of these oxides are found in silicate minerals.

In silicates it is possible to form a variety of structures by altering the bonding of SiO_4 tetrahedra. When all four corners of the SiO_4 tetrahedra are shared with neighboring SiO_4 tetrahedra, the structure has Si:O $= 1:2$ and forms a three-dimensional network structure. Silicates with O/Si $= 2.5$ form a two-dimensional network structure. Those with O/Si $= 3$ form either a one-dimensional chain or ring structures. Silicates with O/Si $= 2.75$ form a double chain structure. These structures are listed in Table 2.5 and are shown in Fig. 2.28. The closer O/Si is to 2, the higher the polymerization. The silicate structures with O/Si $= 4$ are said to have no polymerization.

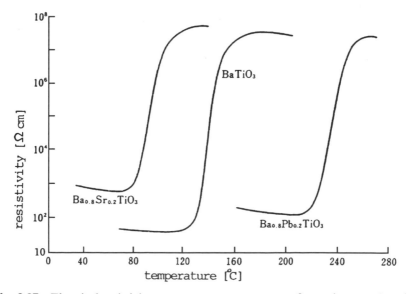

Fig. 2.27 Electrical resistivity versus temperature curves for various semiconducting $BaTiO_3$ ceramics

Table 2.5 Polymerization of SiO_4^{2-} in various silicate minerals

O^{2-}/Si^{4+}	Silicate bases	Structure	Examples
2	SiO_2	3-D network	Quartz, crystobalite
2.5	Si_4O_{10}	2-D network (sheets)	Talc
2.75	Si_4O_{11}	Linear double chains	Asbestos
3	SiO_3	Linear chains	Pyroxenes
		Rings	Beryl
3.5	Si_2O_7	Dimers of SiO_4^{2-}	Akermanite
4	SiO_4	Monomers of SiO_4^{2-}	Olivines

2.3.6.1   MINERALS WITH ISOLATED SILICATES

- Olivine group: Mg_2SiO_4–Fe_2SiO_4 system (Mg^{2+}, Fe^{2+} are both octahedrally coordinated with O^{2-})
- Garnet group: 3 $R''O \cdot R'''_2O_3 \cdot 3SiO_2$ ($R'' = Mg^{2+}$, Fe^{2+}, Ca^{2+}, Mn^{2+}; cubohedrally coordinated, $R''' = Al^{3+}$, Fe^{3+}, Cr^{3+}; octahedrally coordinated)
- Phenakite group: Be_2SiO_4, Zn_2SiO_4 (Be^{2+}, Zn^{2+}; tetrahedrally coordinated)
- Zircon: $ZrSiO_4$ (Zr^{4+}; octahedrally coordinated)

$3CaO \cdot SiO_2 = Ca_3SiO_5$ is a unique compound with isolated silicates and is important as a component of cements due to its large hydration activity. $2CaO \cdot SiO_2 = Ca_2SiO_4$ belongs basically to the olivine group and is a main cement ingredient due to its small hydration activity.

2.3.6.2   MINERALS WITH ISOLATED POLYMERIZED SILICATES

Dimers of silicate bases can be found in the Melilite group which includes $Ca_2MgSi_2O_7$ (Akermanite) and $Ca_2Al(AlSi)O_7$ (Gehlenite). Ca^{2+} in Akermanite is octahedrally coordinated and Mg^{2+} tetrahedrally coordinated.

Benitoite, $BaTiSi_3O_7$, has isolated ring trimers of silicate bases. In the structure both Ba^{2+} and Ti^{4+} are octahedrally coordinated.

Beryl, $Be_3Al_2Si_6O_{18}$, and cordierite, $Al_3Mg_2(Si_5Al)O_{18}$, have hexamers of silicate bases. In beryl Be^{2+} is tetrahedrally coordinated and Al^{3+} octahedrally coordinated. Cordierite has a very low thermal coefficient of expansion and is used widely as honeycomb-shaped catalyst carriers.

2.3.6.3   MINERALS WITH CHAIN SILICATES

The pyroxenes have a general chemical composition of $(SiO_3^{2-})_\infty$ and include $MgSiO_3$, $MnSiO_3$ and $CaSiO_3$. (Mg^{2+} etc. are octahedrally coordinated.)

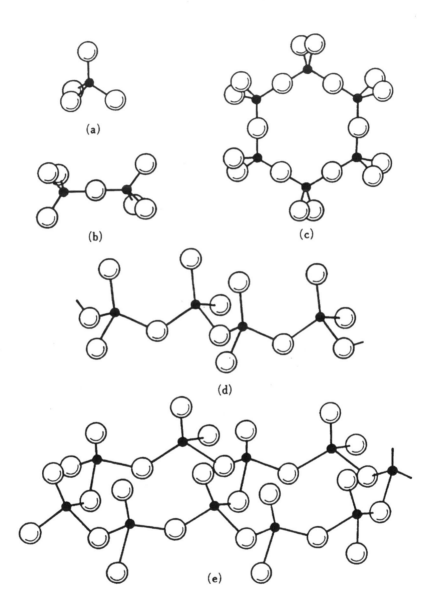

Fig. 2.28   Structures of silicate minerals. (a) O/Si = 4, olivine group with isolated SiO_4;
(b) O/Si = 3.5, Akermanite with dimers of SiO_4; (c) O/Si = 3, beryl with hexagonal rings;
(d) O/Si = 3, garnet with single chains; (e) O/Si = 2.75, asbestos with double chains

The amphiboles have double chains with a general chemical formula of $(Si_4O_{11}^{6-})_\infty$ and include $Ca_2Mg_5(OH)_2(Si_4O_{11})_2$ and $Ca_2(Mg,Fe)_5(OH)_2(Si_4O_{11})_2$. Asbesto, $Mg_6(OH)_6Si_4O_{11} \cdot H_2O$, a well-known fibrous mineral, is a hydrated mineral of tremolite due to weathering.

2.3.6.4   MINERALS WITH PLANAR NETWORKS OF SILICATES

Planar networks are made of $\{(Si_2O_5)H_2\}$ and their arrangements are shown schematically in Fig. 2.29. The layer structures of $Mg(OH)_2$ and $Al(OH)_3$ can also be illustrated similarly (Fig. 2.30). The layer structures of additional minerals with planar networks arc shown in Fig. 2.31.

Minerals without alkali ions between the layers such as talc and kaolinite are very slippery. In addition, kaolinite exhibits plasticity when mixed with water. Minerals with alkali ions between their layers such as mica are easy to cleave.

2.3.6.5   MINERALS WITH THREE-DIMENSIONAL NETWORKS OF SILICATES

The minerals in this class are silica, the feldspars and the zeolites (or sodalites). Silica has many varieties of polymorphs and the stability diagram of these polymorphs is shown in Fig. 2.32. Cristobalite, the stable phase at the highest temperatures, has the diamond or zinc-blende structure. Si ions occupy the carbon positions in the diamond structure and O ions the positions between the two Si ions. In other words, Si ions occupy both the Zn and S positions in the zinc blende structure. Tridymite corresponds to the wurtzite structure. Quartz has a spiral structure and, depending on the direction of the spirals, can be classified as either right-hand or left-hand quartz.

The phase transformations among cristobalite, tridymite and quartz require the reconstruction of the silica networks and thus are called the reconstructive

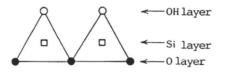

Fig. 2.29   Schematic of a layer structure of silicates

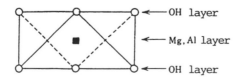

Fig. 2.30   Schematic of a layer structure ($Mg(OH)_2$ or $Al(OH)_3$)

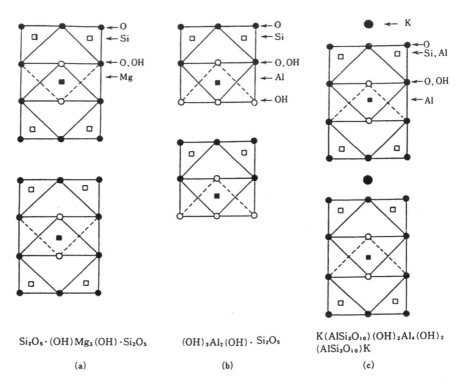

$Si_2O_5 \cdot (OH)Mg_3 (OH) \cdot Si_2O_5$

(a)

$(OH)_3Al_2 (OH) \cdot Si_2O_5$

(b)

$K(AlSi_3O_{10}) (OH)_2Al_4 (OH)_2$
$(AlSi_3O_{10})K$

(c)

Fig. 2.31   Schematics of layer structures of (a) talc, (b) kaolinite and (c) white mica

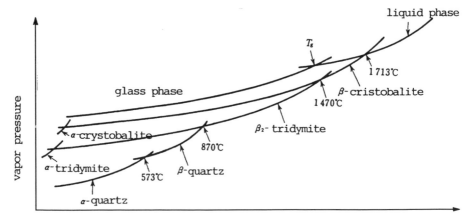

Fig. 2.32   Vapor pressure versus temperature diagram of SiO_2 polymorphs

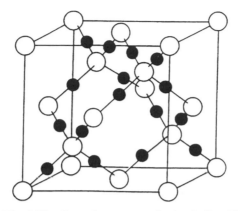

Fig. 2.33 Crystal structure of cristobalite (SiO_2)

transformation. On the other hand, the transformations between the high- and low-temperature phases require a minor movement of ions to adjust Si–O–Si angles and are called the displacive transformation.

In the feldspars a part of Si ions are substituted with Al ions. Alkali and alkaline earth cations are incorporated to satisfy the electrical neutrality of the structures. Because of this incorporation, the feldspars have low melting points. As mentioned previously, porcelains are made of SiO_2 (a skeletal component), a clay such as kaolin (a forming additive) and feldspar (a sintering additive). The feldspars include orthoclase ($KAlSi_3O_8$), albite ($NaAlSi_3O_8$) and anorthite ($CaAl_2Si_2O_8$).

Since the networks are made of SiO_4 rings in zeolites, open channels exist in zeolites, which have been used widely as absorbers, ion exchange media, catalyst carriers and molecular sieves. The zeolites include sodalite ($Na_{16}Al_{16}$ $Si_{24}O_{80} \cdot 16H_2O$) with a channel size of 2.6×3.9 Å, chabazite (Ca_6Al_{12} $Si_{24}O_{72} \cdot 40H_2O$) with a channel size of 3.6×3.7 Å, zeolite-L (K_6Na_3 Al_9 $Si_{27}O_{72} \cdot 21H_2O$) with a channel size of 7.1 Å, and zeolite-A ($Na_{12}Al_{12}$ $Si_{12}O_{48} \cdot 27H_2O$) with a channel size of 4.1 Å.

2.4 STRUCTURES OF GLASSES

Glasses can be defined as supercooled liquids solidified after secondary transformation. When the volume, V, of a glass is measured as a function of temperature, T, V is continuous around a glass transition temperature, T_g, but dV/dT is discontinuous. In the liquid-to-solid transformation V is discontinuous around T_m. The glass transition temperature, T_g, is a function of cooling rate. The higher the cooling rate, the higher the glass transition. (A specific volume increases with increasing cooling rate. This phenomenon has been exploited to strengthen glasses. See Fig. 2.34.)

According to Zachariasen, oxides which form glasses easily have the following structural characteristics;

(1) Each oxide ion should be linked to not more than two cations. Namely, M—O and M—O—M bonds are allowed, but $M_2{=}O{-}M$ or $M_2{=}O{=}M_2$ bonds are not allowed.
(2) The coordination number of oxide ions about the central cations must be equal to or smaller than four.
(3) Oxide ion polyhedra share corners, not edges or faces (see the Pauling rule 3).
(4) At least three corners of each polyhedra should be shared. (this is the requirement for building a three-dimensional network).

Oxide ions having these characteristics are SiO_2, B_2O_3 and P_2O_5. These oxides are called glass formers. Oxides which cannot be a glass former by themselves, but can replace part of the glass-forming oxides, are called intermediates. Those oxides which can lower the polymerization of a glass network and thus can lower the viscosity are called modifiers (Table 2.6). The compositions of several commercial glasses are listed in Table 2.7.

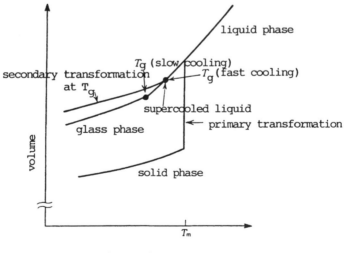

Fig. 2.34   Definition of a glass

Table 2.6 Glass formers, intermediates and modifiers

	Cation	Valence	Coordination number	M–O bond strength (kcal/gatom)	Notes (effects on silicate glasses)
Glass former	B	3	3	119	} Lower softening point, but
			4	89	} do not sacrifice corrosion
	Si	4	4	106	resistance
	Ge	4	4	108	
	P	5	4	$88 \sim 111$	
	V	5	4	$90 \sim 112$	
	As	5	4	$70 \sim 87$	
	Sb	5	4	$68 \sim 85$	
Intermediate	Al	3	4	$79 \sim 101$	Improve corrosion resistance
	Pb	2	2	73	for high refractive index
	Zn	2	2	72	Improve corrosion resistance
	Be	2	4	63	
Modifier	Zn	2	4	36	
	Ba	2	8	33	
	Ca	2	8	32	
	Na	1	6, 8	$15 \sim 25$	Lower softening point for
	K	1	8, 12	$10 \sim 15$	chemical strengthening

Table 2.7 Chemical compositions of some commercial glasses

Components	Soda-lime glass	Lead glass	Borosilicate glass	Boroalumino silicate glass	Zinc borosilicate glass
SiO_2	$65 \sim 75$	$50 \sim 70$	$65 \sim 75$	$65 \sim 70$	$65 \sim 75$
B_2O_3	—	—	$5 \sim 12$	$5 \sim 10$	$5 \sim 10$
Al_2O_3	$0.5 \sim 4$	—	$1 \sim 5$	$5 \sim 10$	$1 \sim 4$
CaO	$5 \sim 15$	$1 \sim 6$	$5 \sim 8$	$3 \sim 7$	$0.5 \sim 1$
ZnO	—	—	<5	—	$5 \sim 10$
PbO	—	$5 \sim 35$	—	—	—
Na_2O	$10 \sim 20$	$2 \sim 15$	$8 \sim 14$	$6 \sim 14$	$7 \sim 10$
K_2O	—	$4 \sim 10$	$1 \sim 5$	$1 \sim 6$	$1 \sim 8$
Notes	Plate glass	High refractive index	Chemical ware		

2.5 STRUCTURAL ANALYSES

2.5.1 Analyses of Crystal Structures

2.5.1.1 CRYSTAL LATTICE

A crystal has a three-dimensional periodic array of atoms and ions. This arrangement of atoms or ions can be represented by the following mathematical expression:

$$\mathbf{R'} = \mathbf{R} + n_1\mathbf{a} + n_2\mathbf{b} + n_3\mathbf{c}$$

where $\mathbf{a}$, $\mathbf{b}$ and $\mathbf{c}$ are three fundamental translation vectors which do not co-exist in a single plane simultaneously and n_1, n_2 and n_3 are arbitrary integers. In a crystal, it is possible to define three vectors, $\mathbf{a}$, $\mathbf{b}$ and $\mathbf{c}$, so that the periodic array seen at a distance, $\mathbf{R}$, from an origin is identical to that seen at $\mathbf{R'}$. In gases it is not possible to define $\mathbf{a}$, $\mathbf{b}$ and $\mathbf{c}$. The minimum space that can be defined by $\mathbf{a}$, $\mathbf{b}$ and $\mathbf{c}$ is called the unit cell. The translational operation, $n_1\mathbf{a} + n_2\mathbf{b} + n_3\mathbf{c}$, fills the space with unit cells. Thus, once the positions of all atoms or ions in a unit cell are identified, it is possible to build a crystal by putting together the unit cells by the translational operation. In other words, it is possible to define a crystal structure by three vectors of the unit cells, $\mathbf{a}$, $\mathbf{b}$ and $\mathbf{c}$, and a set of coordinates for atoms and ions in the unit cell (Fig. 2.35).

As shown in Table 2.8 as well as in Fig. 2.36, there are fourteen kinds of unit cells which can be formed by the translational operation. Those designated by P are called primitive lattices and have an identical arrangement of atoms and ions regardless of the translational operation. In order to express crystal structures with simple and convenient notations of $\mathbf{a}$, $\mathbf{b}$ and $\mathbf{c}$ vectors it is necessary to incorporate additional positions of atoms and ions to the eight corners of the simple lattices. Those having an additional position at the center of the lattices (1/2, 1/2, 1/2) are called body centered lattices and are designated

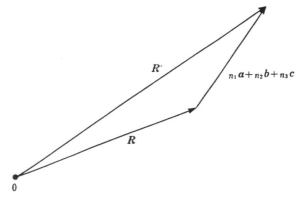

Fig. 2.35 Definition of a crystal by the translational operation

Table 2.8 Fourteen three-dimensional lattices

Crystal group	Number of lattices	Symbol	Constraints on axes and angles
Triclinic	1	P	$a \neq b \neq c$ $\alpha \neq \beta \neq \gamma$
Monoclinic	2	P, C	$a \neq b \neq c$ $\alpha = \gamma = 90° \neq \beta$
Orthorhombic	4	P, C, I, F	$a \neq b \neq c$ $\alpha = \beta = \gamma = 90°$
Tetragonal	2	P, I	$a = b \neq c$ $\alpha = \beta = \gamma = 90°$
Cubic	3	P or sc	$a = b = c$
		I or bcc	$\alpha = \beta = \gamma = 90°$
		F or fcc	
Rhombohedral	1	R	$a = b = c$ $\alpha = \beta = \gamma < 120°$, $\neq 90°$
Hexagonal	1	P	$a = b \neq c$ $\alpha = \beta = 90°$ $\gamma = 120°$

by I. Those having additional positions at the center of faces (0, 1/2, 1/2; 1/2, 0, 1/2; 1/2, 1/2, 0) are called face centered lattices and are designated by F. Those having an additional position at the center of bottom face (1/2, 1/2, 0) are called bottom centered lattices and are designated by C. Three vectors, **a'**, **b'** and **c'**, which connect the nearest-neighbor equivalent positions in I, F and C lattices are the basic translation vectors. For example, a body centered cubic lattice has the following three vectors:

$$\mathbf{a'} = (1/2)\,(x + y - z)\mathbf{a}$$

$$\mathbf{b'} = (1/2)\,(-x + y + z)\mathbf{b}$$

$$\mathbf{c'} = (1/2)\,(x - y + z)\mathbf{c}$$

where x, y and z are the magnitudes of three orthogonal vectors.

   X-ray, electron and neutron beams incident to a crystal with a three-dimensional periodic structure, namely a space lattice, form a three-dimensional periodic diffraction pattern due to diffraction and interference. The space which forms diffraction points is called the reciprocal lattice space or the Fourier space. In this space a reciprocal lattice is formed by a lattice. Vectors in the reciprocal lattice have a dimension of $1/L$. Three basic vectors of a reciprocal lattice, **A**, **B** and **C**, are correlated with three translation vectors of a crystal lattice, **a**, **b** and **c** (**a'**, **b'** and **c'** for I, F, and C space lattices). Their relationships are given by

$$\mathbf{A} = 2\pi(\mathbf{b} \times \mathbf{c})/(\mathbf{a}{\cdot}\mathbf{b} \times \mathbf{c})$$

$$\mathbf{B} = 2\pi(\mathbf{c} \times \mathbf{a})/(\mathbf{a}{\cdot}\mathbf{b} \times \mathbf{c})$$

$$\mathbf{C} = 2\pi(\mathbf{a} \times \mathbf{b})/(\mathbf{a}{\cdot}\mathbf{b} \times \mathbf{c})$$

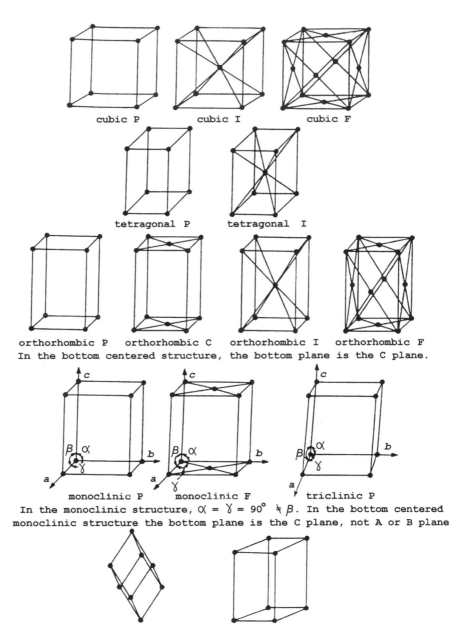

cubic P   cubic I   cubic F

tetragonal P   tetragonal I

orthorhombic P   orthorhombic C   orthorhombic I   orthorhombic F
In the bottom centered structure, the bottom plane is the C plane.

monoclinic P   monoclinic F   triclinic P
In the monoclinic structure, $\alpha = \gamma = 90° \neq \beta$. In the bottom centered
monoclinic structure the bottom plane is the C plane, not A or B plane

hexagonal R   hexagonal P
(rhombohedral)

Fig. 2.36   Fourteen Bravais lattices

The fundamental principle of structural analyses is to transform a diffraction pattern to a crystal lattice by these relationships. The three-dimensional regularity of a diffraction pattern in a reciprocal lattice space is given by

$$\mathbf{G} = h\mathbf{A} + k\mathbf{B} + l\mathbf{C}$$

where h, k and l are the Miller indices. The intensity, F, at a diffraction point, $\mathbf{G}$, is given by

$$F = P \cdot S^* S$$

where P, S and S^* are the correction factor for diffraction measurements, the structure factor and its complex conjugate, respectively.

The structure factor, S, can be given by

$$S(hkl) = \sum f_j \exp[-2\pi i\,(hu_j + kv_j + lw_j)]$$

where u_j, v_j and w_j are the coordinates of the jth atom in the unit cell and f_j is its atomic scattering factor. Furthermore, the atomic scattering factor, f, for X-rays is given by;

$$f = 4\pi \int n(r)r^2[\sin(\mu r)/(\mu r)]dr$$
$$\mu = (4\pi/\lambda)\sin\theta$$

where θ is the Bragg angle, λ the wavelength of the X-ray and $n(r)$ the electron density at a distance, r, from the center of an atom. If all electrons are at the center of an atom, $f = Z$, where Z is the number of electrons in the atom. Also at $\theta = 0$, $f = Z$. But in actual atoms $f < Z$ due to broadening of electron distributions. The atomic scattering factors for Si and O^{2-} are shown in Fig. 2.37.

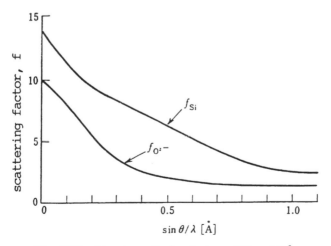

Fig. 2.37   Atomic scattering factors of Si and O^{2-}

Conversely, it is possible to determine the electron distribution of an atom by measuring the θ dependence of f.

2.5.1.2 SYMMETRY OPERATIONS

Two kinds of symmetry operation are possible in crystals. The first is rotation around an axis, by which identical structures can be obtained by the rotation of an angle, $360°/n$, where $n = 1, 2, 3, 4$ and 6. The other is rotation-inversion (or roto-inversion) around an axis, by which identical structures can be obtained first by the rotation at angles, $360°/n$, and then by the inversion. This rotation–inversion can be expressed as $\bar{1}, \bar{2}, \bar{3}, \bar{4}$ and $\bar{6}$. Stereographic projection has been used to indicate the results of the symmetry operations. As shown in Fig. 2.38, the method locates a point N″ on the equatorial plane by first connecting a point N which corresponds to a plane with Miller indices, (hkl), in the northern hemisphere to an origin, extrapolating the line to a point N′ on the sphere, and connecting the point N′ to the south pole of the sphere. The point N″ is the intersection of this connecting line with the equatorial plane. When the line connecting the origin O to the point N is perpendicular to a (hkl) plane, the point N′ represents all planes which are parallel to the (hkl) plane. If a (hkl) plane has n-fold rotational symmetry around the N_o to S_o axis, its projection onto the equatorial plane also has identical n-fold rotation symmetry. In general, patterns in the northern and southern hemispheres are distinguished by o and x, respectively. Figure 2.39 shows all stereographic projections possible by the rotation and by rotation–inversion.

The twofold symmetry of the rotation–inversion can be expressed by m because of the presence of mirror reflections. On the other hand, the sixfold symmetry of the rotation–inversion can be expressed by a combination of a threefold rotation and a mirror reflection at the equatorial plane. In the accepted convention the north-to-south pole axis is the principal axis of symmetry if there is only one symmetry axis. If there are two or more orthogonal symmetry axes, the symmetry axis with the highest n is selected to

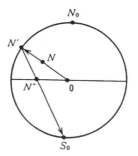

Fig. 2.38 Principle of stereo projection

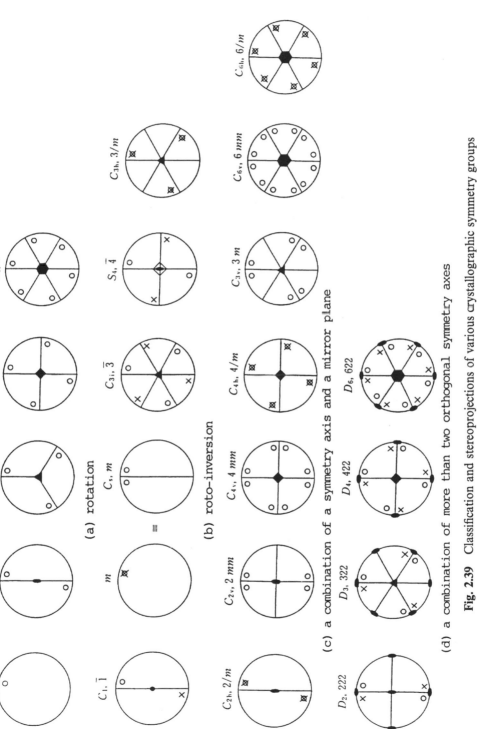

Fig. 2.39 Classification and stereoprojections of various crystallographic symmetry groups

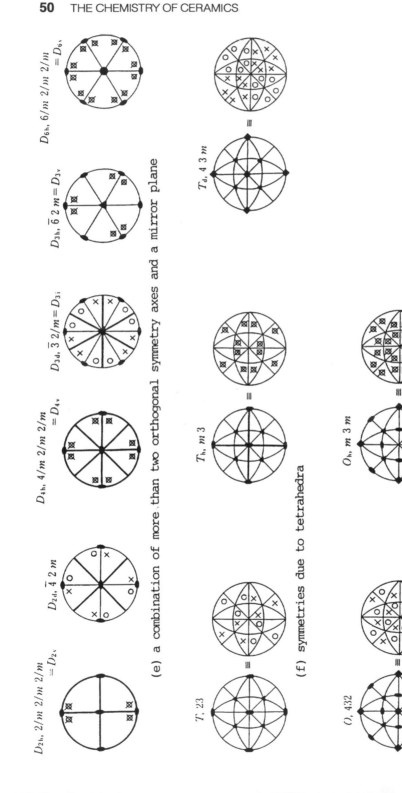

(e) a combination of more than two orthogonal symmetry axes and a mirror plane

(f) symmetries due to tetrahedra

(g) symmetries due to octahedra

Fig. 2.39 (*continued*) Classification and stereoprojections of various crystallographic symmetry groups

be the north-to-south pole axis. A symmetry group with only one axis of symmetry is designated *C*. A symmetry group with two or more orthogonal axes of symmetry is designated *D*. A symmetry group with a threefold rotational axis which does not intersect diagonally with a two- or fourfold rotational axis is designated either *T* (tetrahedron) if the symmetry arises due to the presence of tetrahedra or *O* (octahedron) if the symmetry arises due to the presence of octahedra. These designations for a mirror plane which contains an axis of symmetry are subscripted with v (vertical). When the equatorial plane is the mirror plane, the designations are subscripted with h (horizontal). There are 32 symmetry groups which can be generated by the rotation, rotation–inversion and translation operations and these are shown in Fig. 2.39. In the figure the subscript d denotes diagonal.

Some physical properties are unique to the symmetry groups. The relationships between the physical properties and the symmetry groups are listed in Table 2.9. $BaTiO_3$ belongs to the *O* symmetry group at temperatures above 120°C and to the *C* symmetry group below 120°C. Piezoelectric and ferroelectric properties are present in $BaTiO_3$ only at temperatures below 120°C.

2.5.1.3 STRUCTURAL ANALYSIS METHODS

Analyses of crystal structures are carried out by Fourier transforms of diffraction patterns (reciprocal lattices) which are generated by X-ray, electron and neutron beams. Incident rays or beams are scattered by every lattice point in the crystal, but due to interference of the scattered rays (or beams), images are formed only at certain points in the reciprocal lattice space (the formation of a reciprocal lattice) as shown in Fig. 2.40. Some characteristics of X-ray, electron and neutron beams are summarized in Table 2.10. X-ray techniques

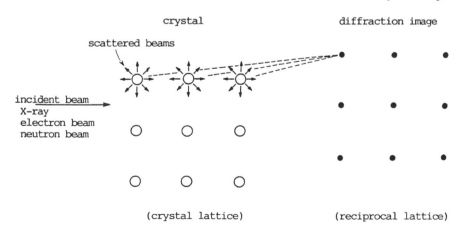

Fig. 2.40 Formation of a diffraction pattern in a reciprocal lattice by the scattering of an incident beam by a crystal lattice

Table 2.9 Crystal symmetry groups and the presence of some physical properties

	Symmetry group		Rotatory power	Left–right images	Piezo-electricity	Pyro-electricity
No.	Notation	Symmetry operation				
1	C_1	1	O	O	O	O
2	C_2	2	O	O	O	O
3	C_3	3	O	O	O	O
4	C_4	4	O	O	O	O
5	C_6	6	O	O	O	O
6	C_1	$\bar{1}$	—	—	—	—
7	C_s	m	O	—	O	O
8	C_{3i}	$\bar{3}$	—	—	—	—
9	C_4	$\bar{4}$	O	—	O	—
10	C_{3h}	$3/m, \bar{6}$	—	—	O	—
11	C_{2h}	$2/m$	—	—	—	—
12	C_{2v}	$2mm$	O	—	O	O
13	C_{3v}	$3m$	—	—	O	O
14	C_{4v}	$4mm$	—	—	O	O
15	C_{4h}	$4/m$	—	—	—	—
16	C_{6v}	$6mm$	—	—	O	O
17	C_{6h}	$6/m$	—	—	—	—
18	D_2	222	O	O	O	O
19	D_3	322	O	O	O	—
20	D_4	422	O	O	O	—
21	D_6	622	O	O	O	—
22	D_{2h}	$2/m\ 2/m\ 2/m$	—	—	—	—
23	D_{2d}	$\bar{4}\ 2m$	O	—	O	—
24	D_{4h}	$4/m\ 2/m\ 2/m$	—	—	—	—
25	D_{3d}	$\bar{3}\ 2/m$	—	—	—	—
26	D_{3h}	$\bar{6}\ 2m$	—	—	O	—
27	D_{6h}	$6/m\ 2/m\ 2/m$	—	—	—	—
28	T	23	O	O	O	—
29	T_h	$m\ 3$	—	—	—	—
30	T_d	$4\ 3m$	—	—	O	—
31	O	432	O	O	—	—
32	O_h	$m\ 3m$	—	—	—	—

O: present; —: not present.

are the most commonly used for structural analyses and several X-ray techniques will be discussed below.

The Generation of X-rays

X-rays are produced when a beam of electrons, emitted from a filament and accelerated by a high voltage, strikes a metallic cathode and loses its kinetic

Table 2.10 Characteristics of X-ray, electron and neutron beams

	Mechanism of scattering and diffraction	Merits	Demerits
X-ray ($\sim$Å)	Weak interaction with X-ray and electrons in a crystal X-rays have wavelengths which are same order of magnitude with atomic distances	Because of their good depths of penetration, X-rays can be used for many crystals Except X-rays with very short wavelengths, it is possible to use X-rays in air	Because of the weak interaction between X-ray and electrons, it is not possible to obtain diffraction patterns from either small or thin crystals. In general, it is necessary to have crystals larger than 100 Å X-rays have very weak scattering from light elements
Electron beams (<0.01 Å)	Strong interaction of electron beams with electrons in a crystal. Because the wavelengths of electron beams are much shorter than atomic distances, it can be assumed that $\sin\theta \sim \theta$ for the Bragg diffraction condition, $2d\sin\theta = n\lambda$	It is possible to do structural analyses of thin and small crystals (~ 10 Å). Electron beams have stronger diffraction from lighter elements than X-rays	It is necessary to have high-voltage electron beams for the analyses of thick samples. It is also necessary to have a high vacuum
Neutron beams ($\sim$Å)	Interaction of magnetic moments of neutrons with those in a crystal	It is possible to do structural analyses of magnetic crystals	It is necessary to have a neutron source

energy either by ejecting electrons from the inner shells of the atoms or by decelerating by collision with the atoms. The cathode is usually made of a high thermal conductivity metal such as Cu, Ag, Fe or Mo.

X-rays generated due to the transitions of electrons among the electron shells are characteristic of a specific atom and are called characteristic X-rays. The transitions to principal quantum numbers, $n = 1, 2, 3$, and 4, are designated K, L, M, and N, respectively. The transitions with $\Delta n = 1, 2, 3$, and 4 are designated α, β, γ, and δ, respectively. The K_α line of copper, CuK_α, has been most widely used for structural analyses. Its wavelength is 1.541 Å.

X-rays generated by the deceleration of incident electrons have a wide and continuous spectrum and are called continuous. The continuous X-ray can be eliminated from the characteristic X-ray either by a filter, a substance which absorbs X-rays with wavelengths shorter than the characteristic X-ray (for example, Ni for CuK_α), or by the monochromatization by Bragg reflection using a crystal such as graphite or mica which has a layer structure and can be bent easily (monochrometer).

The Laue Method

A crystal is irradiated with a continuous X-ray. Lattice spacings of the crystal are discontinuous but the X-ray wavelengths are continuous. Thus it is possible to form a diffraction pattern which will satisfy the Bragg reflection rule (Fig. 2.41). Since the diffraction pattern corresponds to a stereographic projection, it is possible to obtain a fundamental understanding of crystal symmetry by taking several photographs with various angles of rotation.

The Rotating Crystal Method

The Laue method allows only the determination of major symmetry. For more precise analyses of crystal structures, it is necessary to employ one of the following; (1) the rotating crystal method, (2) the precession method, or (3) the Weissenberg camera method. The principle of the rotating crystal method is shown in Fig. 2.42. Because of the rotation of a crystal at the center of a camera, a number of lattice planes satisfy the Bragg reflection rule even for a characteristic X-ray and thus form a diffraction pattern on a film. Since the intensity of each diffraction point is determined by the structure factor for a given (hkl) plane, it is possible to analyze structures in detail measuring the intensities of a large number of diffraction points. There are many diffraction points, around 10^4, for some complicated crystals.

The rotating crystal method allows crystals to rotate 360°. The precession method permits the rotation of crystals within a limited angle. The

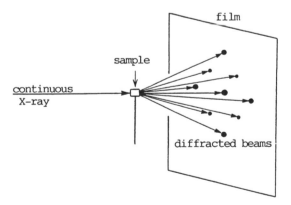

Fig. 2.41 Determination of a crystal's symmetry by the Laue method

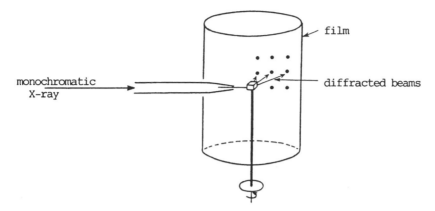

Fig. 2.42 Principle of the rotating crystal method

Weissenberg camera method allows the precession of crystals together with the synchronized motion of an X-ray film. In these two methods each diffraction point corresponds to a single (*hkl*) plane and thus they are better suited for the analyses of structures than the rotating crystal method. It is possible to observe all diffraction points by changing the axes of rotation or precession with respect to the symmetry axes of a crystal.

The Powder Method

When it is not possible to obtain a single crystal of a sample of interest, the powder method can be used to determine its structure. A powder sample corresponds to time-averaged diffraction measurements when the crystal in the rotating crystal method is rotated three-dimensionally. Since each crystal possesses a set of lattice spacing $\{d_1 \ldots d_n\}$ which corresponds to (*hkl*) planes in the crystal, diffraction peaks are observed at angles $\{\theta_1 \ldots \theta_n\}$ which satisfy the Bragg reflection rule. In general, a detection counter is rotated twice more than the angle of rotation of a powder sample (Fig. 2.43). Examples of powder diffraction patterns are shown in Fig. 2.44. Major characteristics of the powder method are as follows:

(1) It is possible to analyze structures of samples even when their single crystals cannot be obtained.
(2) Diffraction peaks from (*hkl*) planes which have an identical d spacing are superimposed upon each other at an angle, θ (for example, (333) and (511) planes of a cubic crystal).
(3) Samples with fibrous or plate-like morphologies tend to exhibit preferred orientations.
(4) Since the power method is simpler and less time consuming than the

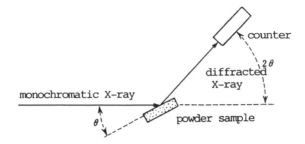

Fig. 2.43 Principle of the powder X-ray diffraction method

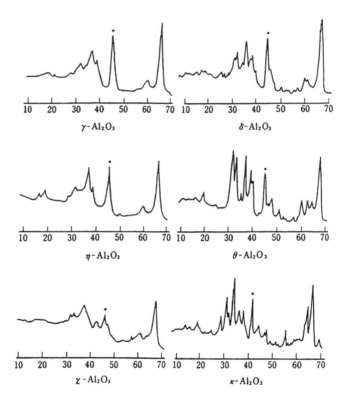

Fig. 2.44 Identification of various Al_2O_3 phases by the powder X-ray method

rotating crystal method, it is possible to carry out an easy determination of crystal phases. (There are many lattice spacing (d) versus intensity (I) data for a variety of substances. Thus it is possible to determine the identity of a sample by matching its d versus I data with those of a known substance.)

(5) It is possible to determine quantitatively the presence of two or more species in a powder. (This technique has been used for the control of firing processes in cement factories.)

(6) It is possible to estimate the sizes of crystallite from the broadening of the diffraction peaks, which is usually expressed in half-height widths. Once the half-height widths, β[rad], are compensated for broadening introduced by the diffractometer, the size of crystallites, D, can be given by

$$D = 0.9\lambda/\beta\cos\theta$$

where λ is the X-ray wavelength. By this method it is possible to measure crystallite sizes between 50 and 1000 Å. It is also possible to determine strains in crystals from the 2θ dependence of the half-height width. Other methods used to determine the sizes of crystallites include electron microscopy and BET adsorption isotherm methods. Since the above three methods give three different crystallite sizes, it is highly recommended to employ all of them.

2.5.2 Structural Analyses of Glasses

It is not possible to perform translational operation on glasses. As a consequence, there are no lattices in glasses and thus we cannot define reciprocal lattices. Unlike gases, glasses are solid and it is necessary to have almost constant distances between the nearest neighbors and also between the next nearest neighbors. This short-range order can be investigated by an analysis of a radial distribution function. The function corresponds to the probability of finding electrons at both ends of a rod, which is rotated at random three-dimensionally, with a distance r. Although the experimental methods used to determine the radial distribution function are similar to the powder method, it is necessary to have a better monochromatized and better collimated X-ray than that used for the powder method. Experimental spectra, not diffraction patterns, can be converted to radial distribution functions by the Fourier transform. The radial distribution function for $ZnCl_2$ is shown in Fig. 2.45. In this glass the nearest neighbors are Zn and Cl ions. Zn ions are coordinated with four Cl ions. The next nearest neighbors are between two Zn ions as well as between two Cl ions. Zn ions are coordinated with four Zn ions and Cl ions with eight Cl ions.

It is also possible to determine liquid structures by analysis of the radial distribution function. For example, Ar atoms in liquid Ar are coordinated with 10 to 11 Ar atoms. (The close packing of solid spheres should give a coordination number of 12.)

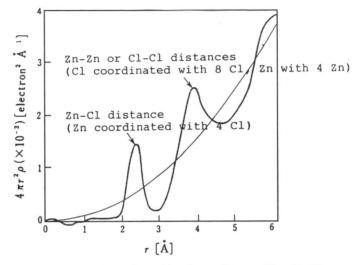

Fig. 2.45 Radial distribution of $ZnCl_2$ glass. (From Hiroshi Hasegawa, PhD Dissertation, Tokyo University, 1979, p. 152)

2.6 LATTICE DEFECTS

2.6.1 Classification of Lattice Defects

So far we have discussed ideal structures which do not contain any defects. At absolute zero temperature crystals are assumed to be most stable and defect-free, but this is simply a thermodynamic speculation. There are a number of defects in crystals. These are called lattice defects simply because they are disturbing three-dimensional regularity of atomic or ionic arrangements in an ideal crystal. Lattice defects play significant roles in solid-state reactions and electrical properties.

Various lattice defects are summarized in Table 2.11. Electronic defects are created by the formation of non-stoichiometric compounds and solid solutions. Point defects are shown in Fig. 2.46. They are a vacancy (a void at a regular lattice site), an interstitial atom (an atom located at an interstitial site of a regular lattice), an impurity atom, and a substitutional atom (a cation substituted with an anion or vice versa). An associated center is a group of ions gathered together by Coulombic force, which is usually isolated from the rest of a crystal. A linear array of lattice defects is called dislocation and a three-dimensional assembly of periodic lattice defects caused by a strong interaction force is called either the superlattice, the crystallographically sheared structure or the cluster. These defects are called extended defects and are often observed in nonstoichiometric compounds with very high defect concentrations.

Table 2.11 Classification of lattice defects

Defects	Species
Electronic defects	Electron
	Electron hole
Point defects	Vacancy
	Interstitial atom
	Substitutional atom
	Impurity atom
	Associated center
Extended defects	Cluster
	Crystallographically sheared structure
	Block structure
Line defects	Dislocation
Plane defects	Surface
	Grain boundary

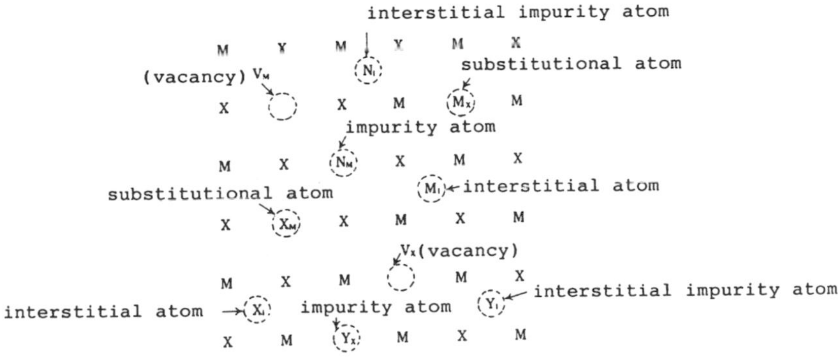

Fig. 2.46 Basic point defects of a MX crystal

Surfaces and grain boundaries are planes where the three-dimensional regularity of a crystal is interrupted. Thus they can also be viewed as (planar) defects.

2.6.2 Point Defects

The Kroger–Vink notation has been used widely to denote point defects. Thus the symbols and system used by Kroger and Vink will also be employed in this book. According to the notation, a defect, A, with an effective charge, b, at a site, a, can be indicated as follows:

$$A_a^b \begin{cases} A: & \text{atomic symbol, V: vacancy} \\ a: & \text{atomic site, i: interstitial site} \\ b: & \text{effective charge, positive (`), negative ('), neutral (}^x) \end{cases}$$

The electric charge of a defect is indicated as an effective charge with respect to the charge of an original ion. In a crystal, M^+X^-, the vacancies of M and X ions are indicated by V'_M and V'_X, respectively, the interstitials of M and X by $M_i^{\cdot}$ and X'_i, respectively, and the impurities, L^{2+} and Y^{2-}, substituing in M and X lattice sites by $L_M^{\cdot}$ and Y'_X, respectively. When these impurities are substituted at interstitial sites, they are indicated by $L_i^{\cdot\cdot}$ and Y_i'', respectively. Thus the electric charge of a defect is the deviation from the electric charge of an ion at a site, a, and is indicated by $'\cdot^x$. It is difficult to position electronic defects, i.e. electrons and electron holes, precisely and thus they are indicated simply by e' and $h^{\cdot}$, respectively.

There are two kinds of points defects, namely Schottky defects and Frenkel defects in ionic crystals. The Schottky defects are pairs of cation and anion vacancies and the Frenkel defects are pairs of a vacancy and an interstitial ion. For crystals with a composition MX the Schottky defect is denoted by $V'_M-V'_X$ and the Frenkel defect by $V'_M-M_i^{\cdot}$ or $V'_X-X'_i$. As indicated by the Kroger–Vink notation, the sum of effective charges of defects is always zero and thus ionic crystals are maintained electrically neutral.

Consider defect concentrations at equilibrium for an ionic crystal MX with Schottky defects. When the defect concentrations are small and thus their interaction can be neglected, the free energy of the crystal can be given by

$$G = Ng_0 + g_{VM}N_{VM} + q_{VX}N_{VX}$$
$$- kT\ln[N!/(N - N_{VM})!N_{VM}!]$$
$$- kT\ln[N!/(N - N_{VX})!N_{VX}!] \tag{2.1}$$

where N is the number of lattice points of both cations and anions in a mole and g_0 is the free energy of one molecule in a perfect crystal. g_{VM} and g_{VX} are the free energies of formation of V'_M and V'_X, respectively, and contain contributions from other effects such as vibrational entropies. The fourth and fifth terms of Eq. (2.1) are the configurational entropies of both vacancies in lattice points. From the electrical neutrality requirement,

$$N_{VM} = N_{VX} = N_S \tag{2.2}$$

From Sterling's formula for a large value of N,

$$\ln N! \approx N\ln N - N \tag{2.3}$$

By differentiating Eq. (2.1) with respect to N_S,

$$\partial G/\partial N_S = g_{VM} + g_{VX} + 2kT\ln\{N_S/(N - N_S)\} \tag{2.4}$$

At equilibrium, $\partial G/\partial N_S = O$. Thus from Eq. (2.4)

$$N_S/N \approx N_S/(N - N_s) = \exp\{-(g_{VM} + g_{VX})/2kT\} \tag{2.5}$$

Equation (2.5) indicates that the concentration of Schottky defects is a function of temperature alone. Since the enthalpies of vacancy formation are positive, the concentration of Schottky defects increases with increasing temperature. A similar argument can be made for Frenkel defects.

As defect structures become more complicated, it becomes difficult to devise the free energy equation as discussed above. If the formation of defects can be viewed as a kind of chemical reaction, it is possible to apply the mass action law, familiar to chemists. Assuming that the formation of Schottky defects is the separation of both cations and anions from their lattice points and their transport to a surface or grain boundaries where they recombine to form lattices, the chemical reaction can be expressed as

$$M_M^x + X_X^x = V_M' + V_X^\cdot + M_S^x + X_S^x \tag{2.6}$$

where M_S^x and X_S^x are M and X ions at their lattice points at a surface, respectively. Equation (2.6) can be simplified as

$$\text{null} = V_M' + V_x^\cdot \tag{2.7}$$

The free energy of this reaction corresponds to the free energy of the formation of Schottky defects. By applying the mass action law with an equilibrium constant of K_S,

$$[V_M'] [V_x^\cdot] = K_S = \exp\{-(g_{VM} + g_{VX})/kT\} \tag{2.8}$$

where $[V_M'] = N_{VM}/N$ and $[V_X^\cdot] = N_{VX}/N$.

From the electrical neutrality requirement,

$$[V_M'] = [V_X^\cdot] = N_S/N$$

Thus Eq. (2.8) can be simplified as

$$N_S/N = \exp\{-(g_{VM} + g_{VX})/2kT\} \tag{2.9}$$

Equation (2.9) is identical to Eq. (2.5). The formation of Frenkel defects can be expressed as

$$M_M^x = M_i^\cdot + V_M' \tag{2.10}$$

Schottky defects have been observed in ionic crystals such as NaCl and MgO. On the other hand, Frenkel defects have been found in crystals with relatively large separation of lattice points such as CaF_2 and Y_2O_3 and in those with large cationic polarization such as AgBr and AgCl. Several defect reactions and their defect-formation enthalpies are listed in Table 2.12.

2.6.3   Non-stoichiometry

Chemical compositions of many compounds can be conveniently expressed by simple ratios of constituent atoms and ions. But the ratios of transition metal and rare earth metal compounds cannot be expressed by simple integer ratios. These are called either non-stoichiometric compounds or uncertain ratio compounds. Examples of these compounds are $Co_{1-\delta}O$, $Zn_{1+\delta}O$ and $UO_{2+\delta}$ where δ indicates the deviation from a stoichiometric composition.

2.6.3.1   CLASSIFICATION OF NON-STOICHIOMETRIC COMPOUNDS

Chemical compositions of non-stoichiometric compounds are determined by temperature and partial pressure of surrounding gas such as O_2 and S_2. Since the defect formation is controlled by the incorporation and discharge of gaseous molecules, the value of δ is determined by the free energy for the incorporation of a defect in a crystal structure. The values of δ for several non-stoichiometric compounds are listed in Table 2.13.

There are four kinds of non-stoichiometric compound, namely cation deficient, cation excess, anion deficient and anion excess. $Co_{1-\delta}O$, $Fe_{1-\delta}O$,

Table 2.12   Formation enthalpies of defects (eV)

Compound	Defect reaction	Formation enthalpy	Compound	Defect reaction	Formation enthalpy
NaCl	$null = V'_{Na} + V^{\cdot}_{Cl}$	2.2–2.4	MgO	$null = V''_{Mg} + V^{\cdot\cdot}_O$	−6
LiF	$null = V'_{Li} + V^{\cdot}_F$	2.4–2.7	CaO	$null = V''_{Ca} + V^{\cdot\cdot}_O$	−6
CaF$_2$	$F^x_F = V^{\cdot}_F + F'_i$	2.3–2.8	UO$_2$	$O^x_O = V^{\cdot\cdot}_O + O''_i$	3.0
	$Ca^x_{Ca} = V''_{Ca} + Ca^{\cdot\cdot}_i$	−7		$U^x_U = V''''_U + U^{\cdot\cdot\cdot\cdot}_i$	−9.5
	$null = V''_{Ca} + 2V^{\cdot}_F$	−5.5		$null = V''''_U + 2V^{\cdot\cdot}_O$	−6.4
BeO	$null = V''_{Be} + V^{\cdot\cdot}_O$	−6			

Table 2.13   Deviation from stoichiometry, δ of various non-stoichiometric compounds

Oxide	δ	Oxide	δ
Fe$_{1-\delta}$O	0.05–0.15 (1300 K)	Co$_{1-\delta}$O	0–0.01
UO$_{2+\delta}$	0–0.24 (1400 K)	Ni$_{1-\delta}$O	0–0.001
PrO$_{2-\delta}$	0–0.3	Zn$_{1+\delta}$O	$0–10^{-6}$
Mn$_{1-\delta}$O	0–0.1		

$Ni_{1-\delta}O$ and $Cu_{2-\delta}O$ are cation-deficient compounds which have cation vacancies. $Zn_{1+\delta}O$, $Cr_{2+\delta}O_3$ and $Cd_{1+\delta}O$ are cation-excess compounds which have cation interstitials. $PrO_{2-\delta}$ and $ZrO_{2-\delta}$ are anion-deficient compounds which have anion vacancies. $UO_{2+\delta}$ is the anion-excess compound which has anion interstitials. Defect species can be determined by a combination of experimental measurements such as mass density measurements, wet chemical analysis, X-ray diffraction, neutron diffraction, thermogravimetric measurement, electrical conductivity and mass diffusivity. Recently electron microscopy has been used to observe defects directly.

2.6.3.2 CATION-DEFICIENT NON-STOICHIOMETRIC COMPOUNDS

When the defect concentration of a non-stoichiometric compound is small, i.e. δ is small, the defects can be viewed isolated from each other as a first-order approximation. Let us choose $Co_{1-\delta}O$ as an example. $Co_{1-\delta}O$ can be regarded as CoO which incorporates oxygen in its NaCl structure. The defect formation reaction can be expressed as

$$1/2O_2 = V_{Co}^x + O_O^x \tag{2.11}$$

where O_O^x and V_{Co}^x are the oxide ions at their regular oxide ion sites and cation vacancies with two trapped electron holes, respectively. In reality the incorporation of an oxide ion in its regular lattice site forms two Co^{3+} ions to satisfy the electrical neutrality requirement, but according to the Kroger and Vink notation the defect formation reaction is expressed by the formation of cation vacancies with two trapped holes. Co^{3+} can be considered as Co^{2+} with an additional electron hole, but the electron holes are not localized to any given Co^{2+} and are capable of moving around in the crystal. This situation can be expressed by

$$V_{Co}^x = V_{Co}' + h^{\cdot} \tag{2.12}$$
$$V_{Co}' = V_{Co}'' + h^{\cdot} \tag{2.13}$$

Assuming that the equilibrium constants for Eqs (2.11), (2.12) and (2.13) are K_1, K_2 and K_3, respectively, the mass action law gives the following relationships:

$$[V_{Co}^x] = K_1 P_{O2}^{1/2} \tag{2.14}$$
$$[V_{Co}']p = K_2[V_{Co}^x] \tag{2.15}$$
$$[V_{Co}'']p = K_3[V_{Co}'] \tag{2.16}$$

where P_{O2} is the oxygen partial pressure, [ ] is the concentration of the defects indicated and p is the concentration of electron holes. The concentration of O_O^x

is not affected strongly by the formation of the defects and is approximated by unity. δ is given by

$$\delta = [V_{Co}^x] + [V_{Co}'] + [V_{Co}'']$$

Thus δ is determined by the equilibrium constants and oxygen partial pressure. When V_{Co}^x are the majority defects, δ is proportional to $P_{O_2}^{1/2}$ at a given temperature.

When V_{Co}' are the majority defects, Eqs (2.14) and (2.15) give

$$[V_{Co}']p = K_1 K_2 P_{O_2}^{1/2} \tag{2.17}$$

Since $p \approx [V_{Co}']$, Eq. (2.17) gives

$$\delta = [V_{Co}'] = p = (K_1 K_2)^{1/2} P_{O_2}^{1/4} \tag{2.18}$$

Thus both δ and p are proportional to $P_{O_2}^{1/4}$.

When V_{Co}'' are the majority defects, Eqs (2.16) and (2.17) give

$$[V_{Co}'']p^2 = K_1 K_2 K_3 P_{O_2}^{1/2} \tag{2.19}$$

Since $p \approx 2[V_{Co}'']$, Eq. (2.19) gives

$$\delta = [V_{Co}''] = 1/2p = [(1/4)K_1 K_2 K_3]^{1/3} P_{O_2}^{1/6} \tag{2.20}$$

Thus both δ and p are proportional to $P_{O_2}^{1/6}$. Figure 2.47 shows the electrical conductivity of $Co_{1-\delta}O$ as a function of P_{O_2}. The electrical conductivity, σ, is given by

$$\sigma = pe\mu \tag{2.21}$$

where μ and e are the mobility of charge carriers and the electron charge, respectively. Where the mobility of electron holes does not depend on the concentration of electron holes, $\sigma \propto p$. Thus the P_{O_2} dependence of σ is identical to that of the defect concentrations. Figure 2.47 indicates two regimes of P_{O_2} dependence, namely $\delta \propto P_{O_2}^{1/4}$ and $\delta \propto P_{O_2}^{1/6}$, which suggests that the majority defects of $Co_{1-\delta}O$ are V_{Co}' at high P_{O_2} and V_{Co}'' at low P_{O_2}.

CoO is a p-type semiconductor. Its electrical conductivity increases with increasing temperature and oxygen partial pressure. In the band structure of CoO the $2p$ band of oxygen forms a valence band and the $4s$ band of Co forms a conduction band. Non-stoichiometric defects create impurity levels in the forbidden band. As indicated in Eqs (2.12) and (2.13), both V_{Co}^x and V_{Co}' act as acceptors, namely to excite electrons from the valence band and leave electron holes in the valence band. The enthalpy of the defect formation reaction corresponds to the ionization energy which can be viewed as the difference in energy between the acceptor level and the edge of the valence band. The band structure of CoO is shown in Fig. 2.48.

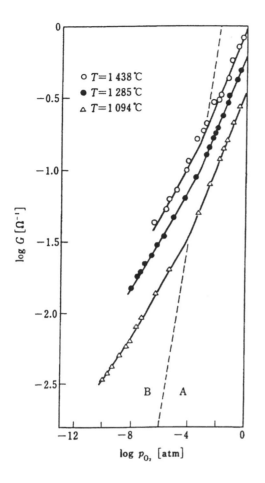

Fig. 2.47 Electrical conductivity versus oxygen partial pressure curves of CoO. (Reproduced by permission of The American Institute of Physics, from B. Fisher and D. S. Tannhauser, *J. Chem. Phys.*, **44**, 1663, 1966)

2.6.3.3 NON-STOICHIOMETRIC COMPOUNDS WITH EXCESS CATIONS

$Zn_{1+\delta}O$ is a non-stoichiometric compound with excess cations and can be viewed as ZnO from which oxide ions are removed because of the equilibrium reaction with an oxygen atmosphere. Its defect equilibrium can be expressed by

$$ZnO = Zn_i^x + (1/2)O_2(g) \tag{2.22}$$

Zn_i^x denotes an interstitial zinc atom which can ionize and donate electrons to the conduction band as follows:

$$Zn_i^x = Zn_i^. + e' \tag{2.23}$$
$$Zn_i^. = Zn_i^{..} + e' \tag{2.24}$$

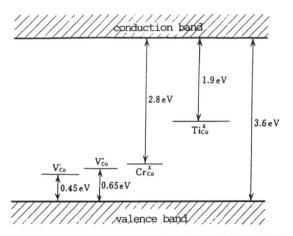

Fig. 2.48 Band structure of CoO. (Reprinted from M. Gvishi and D. S. Tannhauser, *J. Phys. Chem. Solids*, **33**, 893, 1972 with kind permission from Elsevier Science Ltd, The Boulevard, Langford Lane, Kidlington OX5 1GB, UK)

Both Zn_i^x and $Zn_i^{\cdot}$ act as donors and thus ZnO is an *n*-type semiconductor. By applying the mass action law as in the case of $Co_{1-\delta}O$, the following relations can be obtained:

$$\delta \propto P_{O_2}^{-1/2}$$

where Zn_i^x are the majority defects,

$$\delta = [Zn_i^{\cdot}] = n \propto P_{O_2}^{-1/4}$$

where $Zn_i^{\cdot}$ are the majority defects, and

$$\delta = [Zn_i^{\cdot\cdot}] = n \propto P_{O_2}^{-1/6}$$

where $Zn_i^{\cdot\cdot}$ are the majority defects. As shown in Fig. 2.49,

$$\sigma \propto n \propto P_{O_2}^{-1/4}$$

Thus it can be concluded that $Zn_i^{\cdot}$ are the majority defects. At higher temperatures,

$$\sigma \propto n \propto P_{O_2}^{-1/6}$$

which indicates that $Zn_i^{\cdot\cdot}$ are the majority defects at higher temperatures. Figure 2.50 shows the band structure of ZnO.

2.6.3.4 NON-STOICHIOMETRIC COMPOUNDS WITH EXTENDED DEFECTS

As indicated above, it is very effective to discuss the formation of defects from the viewpoint of defect equilibrium when the concentration of defects is low

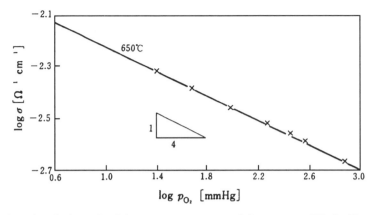

Fig. 2.49 Electrical conductivity versus oxygen partial pressure of ZnO. (From H. H. von Baumbach and C. Wagner, *Z. Phys. Chem.*, **B22**, 199, 1933. Reproduced by permission of R. Oldenbourg Verlag GmbH)

and thus their interaction can be ignored. With increasing concentration of defects it is no longer possible to disregard the presence of interaction among the defects. Several compounds with large amounts of point defects, which tend to align regularly and form extended defects, will be discussed next.

Cluster Structures

Although they are difficult to distinguish from associated defect centers, point defects which exhibit crystallographic short-range order are called clusters. One of the the well-known cluster structures is the Koch cluster in $Fe_{1-\delta}O$. As shown schematically in Fig. 2.51(a), a unit of Koch clusters consists of 13 vacant octahedral cations sites in the rock salt structure and four tetrahedral

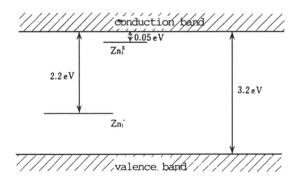

Fig. 2.50 Band structure of ZnO. (From K. Hauffe, *Angew. Chem.*, **72**, 730, 1960: reproduced by permission of VCH)

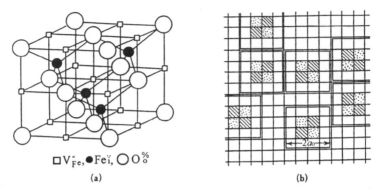

$\Box V_{Fe}^{''}, \bullet Fe_{i}^{...}, \bigcirc O_{o}^{\%}$

(a)   (b)

Fig. 2.51   Koch cluster of $Fe_{1-\delta}O$. (From J. S. Anderson, in C. N. R. Rao (ed.), *Modern Aspects of Solid State Chemistry*, Plenum Press, 1970, p. 29: reproduced by permission of Plenum Press)

interstitial sites occupied by Fe^{3+}. These units form $2 \times 2 \times 2$ arrays and are distributed as shown in Fig. 2.51(b).

$UO_{2+\delta}$ has a fluorite type crystal structure and its excess oxygen ions occupy octahedral interstitial cation sites. The interstitial sites correspond to the position $(1/2, 1/2, 1/2)$ in the unit cell as shown in Fig. 2.52. In reality, these oxygen ions are not located exactly at $(1/2, 1/2, 1/2)$, but displaced about 1 Å along the direction of $\langle 110 \rangle$. Simultaneously, two oxygen ions at the regular lattice sites are displaced about 1 Å along the direction of $\langle 111 \rangle$. These are called Willis clusters.

Crystallographically Sheared Structures

There is a series of Magneli phases with a general formula of Ti_nO_{2n-1} between rutile structure TiO_2 and corundum structure Ti_2O_3. Rutile structure TiO_2 is

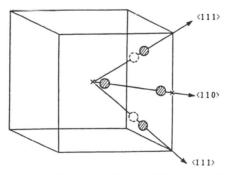

$\langle 111 \rangle$

$\langle 110 \rangle$

$\langle 111 \rangle$

Fig. 2.52   2:2:2 Willis cluster. (From K. H. G. Ashbee, in L. Eyring and M. O'Keefe (eds), *The Chemistry of Extended Defects in Nonmetallic Solids*, North-Holland, 1970, p. 272: reproducd by permission of North-Holland)

known to be an anion-deficient non-stoichiometric compound. With increasing concentration of oxide ion vacancies due to reduction, these oxide ion vacancies are no longer isolated and align along a certain crystallographic plane, which is called sheared. In the rutile structure TiO_6 octahedra share corners along a and b axes and edges along c axes. In the sheared plane TiO_6 octahedra share planes, which permit the decrease in oxide ions necessary for bonding and allow relaxation of crystallographic strains induced by the decrease in oxide ion concentration (see Fig. 2.53). These defect structures are called crystallographically sheared and are found in $(W, Nb)_nO_{2n-1}$, $(W, Ta)_nO_{2n-1}$ and $(W, Mo)_nO_{2n-1}$. The sheared structure of $Cr_2Ti_6O_{15}$ observed by transmission electron microscopy is shown in Fig. 2.54.

2.6.4 Solid Solubility and Point Defects

When one solid dissolves another in a solid state without altering its original crystal structure, these solids form a solid solution. There are two types of solid solution. A substitutional solid solution is formed when the solute atoms occupy regular atomic sites. On the other hand, an interstitial solid solution is formed when solute atoms dissolve into interstitial sites.

2.6.4.1 FACTORS WHICH INFLUENCE SOLID SOLUBILITY

Thermodynamically the solubility of crystal B into crystal A depends on the values of the free energy of a given system. Atomistically the solubility depends on the difference in bonding between A and B, ionic radius, chemical affinity and atomic valence. The solubility of ionic crystals depends especially on ionic

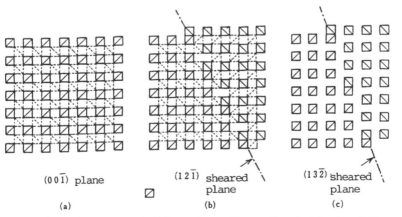

(0 0 $\bar{1}$) plane   (1 2 $\bar{1}$) sheared plane   (1 3 $\bar{2}$) sheared plane

(a)   (b)   (c)

Fig. 2.53 Sheared structure of Ti_nO_{2n-1}. (From Japan Chemical Society (ed.), *Treatise of Chemistry IX: Inorganic Reactions with Solids*, Society Publications Center, 1975, p. 7: reproduced with permission)

Fig. 2.54 Electron micrographs of the sheared structure of $Cr_2Ti_6O_{15}$. (a) Lattice image of $(1\bar{2}1)$, sheared structure; (b) enlargement of (a); (c) corresponding electron diffraction pattern (From S. Kamiya, M. Yoshimura and S. Somiya, *Mat. Res. Bull.*, **15**, 1303, 1980: reproduced by permission of The Materials Research Society)

radius, atomic valence and crystal structure. For example, let us assume the dissolution of CoO, a rock salt structure crystal, into ZnO, a wurtzite structure crystal. There are two possibilities; one is that Co^{2+} is incorporated into the wurtzite structure and substitutes Zn^{2+} with an associated change in coordination number from six to four. Another possibility is that Co^{2+} can occupy octahedrally coordinated interstitial sites. From a structural viewpoint either possibility is very difficult. From the measurements of visible light spectra it is known that CoO dissolves into ZnO substitutionally, but its solubility limit is about 15% at 1000°C. Although their ionic radii are close, 0.71 Å for Zn^{2+} versus 0.74 Å for Co^{2+}, this is a good example of the solid solutions which have a limited solubility due to differences in crystal structure. Since there is no major difference in either ionic radius, atomic valence or crystal structure between NaCl and KCl or between NiO and MgO, these systems form a continuous solid solution.

2.6.4.2 SOLID SOLUTION AND ATOMIC VALENCE

Example 1

In solid solutions with a limited range of solubility the dissolution of solute atoms or ions creates defects in the host crystal. Point defects are created to maintain electrical neutrality in solid solutions between compounds whose cations have fixed atomic valences. For example, the dissolution of $CaCl_2$ into NaCl creates cation vacancies and can be expressed by

$$CaCl_2 \xrightarrow{NaCl} Ca_{Na}^{\cdot} + V_{Na}' + 2Cl_{Cl}^{x}$$

The dissolution of CaO into ZrO_2 creates anion vacancies and can be expressed by

$$CaO \xrightarrow{ZrO_2} Ca_{Zr}'' + V_{O}^{\cdot\cdot} + O_{O}^{x}$$

The dissolution of YF_3 in CaF_2 creates anion interstitials and can be expressed by

$$YF_3 \xrightarrow{CaF_2} Y_{Ca}^{\cdot} + 2F_{F}^{x} + F_{i}'$$

The formation of these point defects has been confirmed mainly by the measurements of bulk densities. In general, the dissolution of higher-valence cations into a host compound with lower valence cations tends to create either cation vacancies or anion interstitials. On the other hand, the dissolution of lower-valence cations into a host compound with higher-valence cations tends to create either anion vacancies or cation interstitials.

Interaction among point defects becomes important with increasing amounts of point defects. The dissolution of CaO into ZrO_2 creates oxygen vacancies as discussed above. The presence of oxygen vacancies makes oxygen ions mobile and thus makes ZrO_2 doped with CaO a good oxygen ion conductor. $[V_{O}^{\cdot\cdot}]$ increases almost linearly with an increasing amount of CaO in ZrO_2. The ionic conductivity of ZrO_2 doped with up to 15 mol% CaO also increases linearly with increasing $[V_{O}^{\cdot\cdot}]$ (Fig. 2.55). Ionic conductivity decreases with increasing concentrations of CaO above 15 mol%. The decrease in ionic conductivity can be attributed to the decreasing concentration of mobile oxygen vacancies due to the formation of associated defects between Ca_{Zr}'' and $V_{O}^{\cdot\cdot}$.

Example 2

The dissolution of cations with a fixed atomic valence into a host compound with variable valence cations will be discussed next. The dissolution of about 1 mol% Al_2O_3 into ZnO forms a substitutional solid solution and its defect formation reaction can be expressed as

$$Al_2O_3 \rightarrow 2Al_{Zn}^{\cdot} + 2O_{O}^{x} + 2e' + (1/2)O_2$$

The concentration of electrons created by the dissolution of Al_2O_3 is significantly larger than the intrinsic concentration of electrons created by

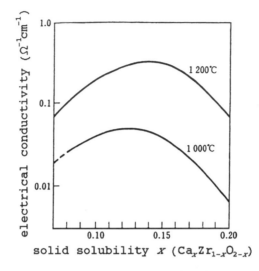

Fig. 2.55 Electrical conductivity as a function of CaO content in CaO-stabilized zirconia

Eqs (2.22) and (2.24). Thus it is possible to assume that the concentration of conduction electrons is almost identical to that of Al^{3+}.

The dissolution of Li_2O into ZnO decreases the concentration of conduction electrons by

$$Li_2O + 2'e + (1/2)O_2 \rightarrow 2Li'_{Zn} + 2O_O^x$$

As discussed above, it is expected that the dissolution of Al_2O_3 into ZnO increases electrical conductivity, but the dissolution of Li_2O decreases it. The experimental data of electrical conductivity, σ, as a function of solute concentrations is shown in Fig. 2.56, which supports the predicted dependencies of solute concentrations on conduction electrons.

Example 3

The dissolution of Li_2O and Al_2O_3 into CoO can be treated similarly and their defect formation reactions are given by

$$Li_2O + (1/2)O_2 \rightarrow 2Li'_{Co} + 2h^\cdot + 2O_O^x$$
$$Al_2O_3 + 2h^\cdot \rightarrow 2Al^\cdot_{Co} + 2O_O^x + (1/2)O_2$$

Thus the dissolution of Li_2O increases the concentration of electron holes, but the dissolution of Al_2O_3 decreases this concentration. In contrast to the solid solutions of ZnO, electrical conductivity increases with the dissolution of

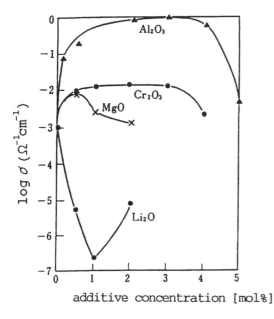

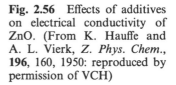

Fig. 2.56 Effects of additives on electrical conductivity of ZnO. (From K. Hauffe and A. L. Vierk, *Z. Phys. Chem.*, **196**, 160, 1950: reproduced by permission of VCH)

lower-valence cations and decreases with the dissolution of higher-valence cations.

The manipulation of valence states of host cations by dissolving higher-valence cations in *n*-type semiconductors and lower-valence cations in *p*-type semiconductors is called controlling valence. Thus semiconductors which form solid solutions with additives are called valence-controlled semiconductors. Examples of valence-controlled semiconductors are listed in Table 2.14. In the table it is expected that both Zn'_{Zn} and Ti'_{Ti} are dissociated into $Zn^x_{Zn} + e'$ and $Ti^x_{Ti} + e'$, respectively.

2.7 DISLOCATIONS

Dislocations are line defects introduced during the deformation of crystals by shear stresses. There are two types of dislocations, namely edge and screw. The former is shown on the right side of Fig. 2.57 and the latter on the left side. Usually both types of dislocation can be observed in real crystals. The areas delineated by arrows in the figure have irregular arrays of atomic arrangement. The centers of the areas are called dislocation cores and the lines connecting dislocation cores are dislocation lines.

Let us make a circuit by connecting atoms around a dislocation core and start at an atom away from the core and count a given number of lattice distances in one direction. Then we count another number of lattice distances

Table 2.14 Valence-controlled semiconductors

Crystal	Additive	Lattice defects formed	Ionic radius Å	Semi-conductor type	Notes
NiO	Li_2O	Li'_{Ni}, $Ni^{\cdot}_{Ni}$	Li^+ (6) = 0.68, Ni^{3+} (6) = 0.60, Ni^{2+} (6) = 0.69	p	Narrow d-orbital band or hopping thermistor
CoO	Li_2O	Li'_{Co}, $Co^{\cdot}_{Co}$	Li^+ (6) = 0.68, Co^{3+} (6) = 0.63, Co^{2+} (6) = 0.72	p	Narrow d-orbital band or hopping thermistor
FeO	Li_2O	Li'_{Fe}, $Fe^{\cdot}_{Fe}$	Li^+ (6) = 0.68, Fe^{3+} (6) = 0.64, Fe^{2+} (6) = 0.74	p	Narrow d-orbital band or hopping thermistor
MnO	Li_2O	Li'_{Mn}, $Mn^{\cdot}_{Mn}$	Li^+ (6) = 0.68, Mn^{3+} (6) = 0.70, Mn^{2+} (6) = 0.80	p	Narrow d-orbital band or hopping thermistor
ZnO	Al_2O_3	$Al^{\cdot}_{Zn}$, Zn'_{Zn}	Al^{3+} (4) = 0.49, Zn^+ (4) = 0.93, Zn^{2+} (4) = 0.71, (0.49 + 0.93)/2 = 0.71	n	sp^3 anti-bond or O^{2-}-band
TiO_2	Ta_2O_5	$Ta^{\cdot\cdot}_{Ti}$, Ti'_{Ti}	Ta^{5+} (6) = 0.68, Ti^{4+} (6) = 0.68, Ti^{3+} (6) = 0.76	n	
Bi_2O_3	BaO	Ba'_{Bi}, $Bi^{\cdot}_{Bi}$	Ba^{2+} (8) = 1.43, Bi^{4+} (8) = 0.85, Bi^{3+} (8) = 1.00	p	High resistance or narrow band (ZNR varistor component)
Cr_2O_3	MgO	Mg'_{Cr}, $Cr^{\cdot}_{Cr}$	Mg^{2+} (6) = 0.66, Cr^{4+} (6) = 0.63, Cr^{3+} (6) = 0.69	p	
Fe_2O_3	TiO_2	$Ti^{\cdot}_{Fe}$, Fe'_{Fe}	Ti^{4+} (6) = 0.68, Fe^{3+} (6) = 0.64, Fe^{2+} (6) = 0.74	n	
$BaTiO_3$	La_2O_3	$La^{\cdot}_{Ba}$, Ti'_{Ti}	La^{3+} (12) = 1.23, Ba^{2+} (12) = 1.47, Ti^{4+} (6) = 0.68, Ti^{3+} (6) = 0.76	n	PTC thermistor
$BaTiO_3$	Ta_2O_5	$Ta^{\cdot}_{Ti}$, Ti'_{Ti}		n	PTC thermistor
$LaCrO_3$	SrO	Sr'_{La}, $Cr^{\cdot}_{Cr}$	Sr^{2+} (12) = 1.25, La^{3+} (12) = 1.23, Cr^{4+} (6) = 0.63, Cr^{3+} (6) = 0.69	p	High-temperature heating element
$LaMnO_3$	SrO	Sr'_{La}, $Mn^{\cdot}_{Mn}$	Sr^{2+} (12) = 1.25, La^{3+} (12) = 1.23, Mn^{4+} (6) = 0.60, Mn^{3+} (6) = 0.70	p	High-temperature heating element
$K_2O \cdot 11\,Fe_2O_3$	TiO_2	$Ti^{\cdot}_{Fe}$, Fe'_{Fe}		n	Mixed ion-electron conductor
SnO_2	Sb_2O_5	$Sb^{\cdot}_{Sn}$, Sn'_{Sn}	Sb^{5+} (6) = 0.62, Sn^{3+} (6) = 0.81, Sn^{4+} (6) = 0.71, (0.62 + 0.81)/2 = 0.715	n	Transparent electrode

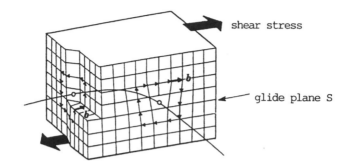

shear stress

glide plane S

Fig. 2.57 Schematics of dislocations

in an orthogonal direction. For lattices with dislocations the end point is one lattice distance away from the starting point. A vector connecting from the end point to the starting point is called a Burger's vector which is commonly designated by **b**. As indicated in Fig. 2.57, the Burger's vector for an edge dislocation is perpendicular to the dislocation line and the Burger's vector for a screw dislocation is parallel to it. The Burger's vector for a mixed dislocation between the edge and screw dislocations forms an angle with the dislocation line. Lattices with edge dislocations glide parallel to the Burger's vector while lattices with screw dislocations glide perpendicular to it.

The Burger's vector is constant for a single dislocation line. The sum of Burger's vectors is also constant even for those dislocation lines which branch into two or more lines. This is called conservation of the Burger's vector. Because of this conservation rule, dislocations form a closed loop within the crystal or terminate at crystal surfaces or grain boundaries.

Dislocations have excess free energies due to lattice distortion and elastic strains around the dislocations cores. Usually the latter contribution is predominant and its excess free energy per unit length can be given by

$$E \approx Gb^2 \qquad \text{(for edge dislocations)}$$
$$E \approx Gb^2/(1 - v) \quad \text{(for screw dislocations)}$$

$$(2.25)$$

where G and v are the shear modulus and Poisson's ratio, respectively. The excess energy, called the dislocation energy, is proportional to b^2 for both types of dislocation. This relationship indicates that dislocations are most stable when their Burger's vectors point in the direction of the highest atomic arrangement. It is known that the Burger's vectors are parallel to $\langle 110 \rangle$ for MgO and $MgAl_2O_4$. Because of a very large Burger's vector, it is possible to grow a single crystal of $Y_3Al_5O_{12}$ without any dislocations.

There are a number of ways to observe dislocations. When the surface of a crystal is etched by a suitable chemical, the areas around dislocations are etched preferentially and form etch pits which can be observed with an optical microscope. For transparent crystals it is possible to observe the segregation of impurities around dislocation lines, which is known as the decoration method. When an electron beam passes through a thin crystal, a contrast is created due to the difference in transmissivity. Thus it is possible to observe dislocations by observing the contrast. Examples are shown in Fig. 2.58.

2.8 STRUCTURES OF SINTERED BODIES

Polycrystalline bodies obtained by either pressureless sintering or hot pressing have complicated microstructures which consist of grains, grain boundaries and pores (see Fig. 2.59). Thus microstructural parameters which control the physical properties of sintered bodies are grain size, grain boundary thickness, size and number of pores, and their geometric distributions. Sintering additives and impurities in starting materials tend to segregate on the surface or along grain boundaries. Thus the amount and distribution of second phases are also important factors which determine the physical properties of the sintered bodies.

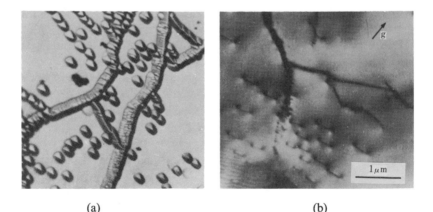

(a)   (b)

Fig. 2.58 Experimental observations of dislocations. (a) Etch pits of TiO_2 (from W. H. Hirthe, N. R. Adsit and J. O. Brittain, in J. B. Newkirk and J. H. Wernick (eds), *Direct Observations of Imperfections in Crystals*, Interscience, 1962, p. 135 ©Interscience); (b) dislocations and small-angle grain boundaries of arc-welded UO_2 (g indicates the (131) direction). (From K. H. G. Ashbee in L. Eyring and M. O'Keefe (eds), *The Chemistry of Extended Defects in Nonmetallic Solids*, North-Holland, 1970, p. 323: reproduced by permission of North-Holland)

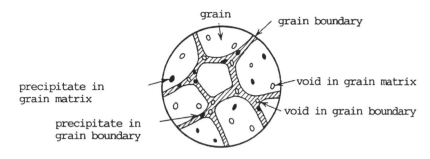

Fig. 2.59 Microstructure of a sintered body

2.8.1 Grain Boundaries

Sintered bodies have grain boundaries which are not present in either single crystals or glasses. Grain boundaries are formed between two or more grains which have different crystallographic orientations and discontinuities in atomic orientation and electronic potential. The thickness of a grain boundary depends on sintering temperature, time, atmosphere and atomic mobility and is between a few tens of Ångstroms and a few micrometers. Sometimes it is proper to consider very thick grain boundaries as boundary layers. A second phase on grain boundaries is called the intergranular phase. General properties of grain boundaries are as follows:

(1) The grain boundary diffusion of atoms and ions is faster than the grain matrix diffusion.
(2) The melting point of grain boundaries is lower than that of the matrix.
(3) Impurities tend to segregate along grain boundaries.
(4) Electrical, magnetic, mechanical and optical properties of grain boundaries are different from those of the matrix.
(5) There is a large number of electron traps in grain boundaries and thus a potential barrier tends to form along the boundaries.

There is no well-established method to determine the composition and structure of a grain boundary and a number of aspects of grain boundaries remain to be investigated in detail. By controlling grain boundaries artificially it is possible to obtain sintered bodies with unique physical properties which cannot be obtained from single crystals. A few examples are as follows:

- *ZnO–Bi$_2$O$_3$ varistors*: The microstructure of ZnO–Bi$_2$O$_3$ varistors consists of ZnO grains, which are low-resistance *n*-type semiconductors, and grain boundary barriers containing Bi$_2$O$_3$. Their microstructure is illustrated schematically in Fig. 2.60(a). The current–voltage characteristic of the varistors is very non-linear as shown in Fig. 2.60(b) and can be given by

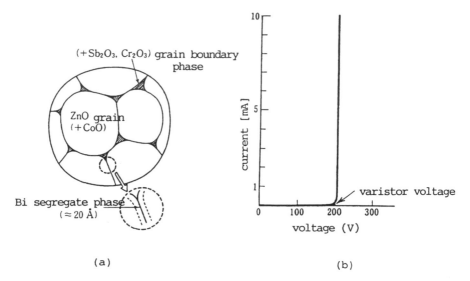

Fig. 2.60   A ZnO–Bi$_2$O$_3$ varistor. (a) Microstructure; (b) current–voltage characteristic

$$I = (V/C)^\alpha$$

where C and α are constants. α for the varistors with additives such as CoO, Sb$_2$O$_3$, MnO and Cr$_2$O$_3$ in addition to Bi$_2$O$_3$ can be as high as 50. Initially it was speculated that the non-linear behavior arises due to dielectric breakdown of the insulating grain boundary phases. Since the varistor voltage E depends not on the thickness of the grain boundaries but on the number of the ZnO grains, it is vital to include the interaction between the grains and grain boundaries for explaining the non-linear behavior. A number of models have been proposed, but the behavior of the varistors has not been explained satisfactorily.

- *BaTiO$_3$ sintered bodies*: When BaTiO$_3$ is doped with a small amount of rare earth oxide additives such as of La or is reduced, BaTiO$_3$ sintered bodies become semiconducting and exhibit low resistance below the Curie temperature but high resistance above it. The electrical resistivity of polycrystalline BaTiO$_3$ bodies is shown in Fig. 2.27. This phenomenon is called the positive temperature coefficient (PTC) effect, which is present only in polycrystalline bodies, but not in the single crystals. In order to exhibit the PTC effect, it is necessary for the grains to be ferroelectric and semiconducting and for the grain boundaries to be insulating.

- *Silicon nitride*: Silicon nitride, Si$_3$N$_4$, has been sought as a high-temperature structural material due to its excellent wear, thermal shock and oxidation resistance characteristics. In general, it is very difficult to sinter Si$_3$N$_4$ without additives. Thus Si_3N_4 is hot pressed with MgO and Y_2O_3 as sintering aids. Mechanical strengths of the resultant Si_3N_4 bodies degrade significantly at temperatures above 1200°C, which has been attributed to the presence of a glassy phase along the grain boundaries. The viscosity of the glassy phase decreases at high temperatures, which makes sliding among the grains easy. As shown in Fig. 2.61, the presence of the glassy phase has been confirmed experimentally.

2.8.2 Pores

There are two types of pore in sintered bodies, open, which are open to the surfaces of the sintered bodies, and closed, which are isolated and not open to a surface. In an initial stage of sintering all pores are open pores. As sintering proceeds open pores decrease and closed pores increase. The amount of pores is called porosity and is given by weight percentage. Naturally, the porosity can be distinguished into either open or closed. The sum of both porosities constitutes the total porosity. Since open pores are open to air, the composition of gases in open pores is identical to that of the ambient air. On the other hand, closed pores contain gases in a sintering atmosphere and gases due to decomposition and evaporation of sintering materials.

The presence of pores tends to be detrimental to functions of sintered bodies. However, it is possible to control pores so that the presence of pores can be

Fig. 2.61 A triple grain boundary. A is a glassy phase. the lower-left corner is out of contrast, but seems to be crystalline from the moiré pattern at B. (From D. R. Clarke and E. Thomas, *Am. Ceram. Soc. Bull.*, **60**, 491, 1977. Reproduced by permission of The American Ceramic Society)

exploited positively in applications such as porous bodies. Negative effects due to the presence of pores are (1) an increase in electrical resistance around pores, (2) a decrease in mechanical strength, (3) an increase in optical scattering, and (4) a decrease in thermal conductivity.

Progress in sintering technology has paved the way to making sintered bodies with very small porosities. It is possible today to sinter polycrystalline bodies which have almost no porosity. For example, translucent polycrystalline bodies have been sintered to almost zero porosity so that the optical scattering due to pores has been eliminated.

2.8.3 Crystalline Grains

As discussed in Section 2.6, it is critical to recognize the presence of lattice defects when crystals are viewed atomistically. In macrostructures such as the structures of polycrystalline bodies, grain sizes and their distribution are two critical factors. When grain growth is suppressed during sintering, sintered bodies with very minute grains can be obtained. On the other hand, it is possible to see the growth of small number of large grains without the control of grain growth, which is called abnormal grain growth and is shown in Fig. 2.62.

In some applications the crystallographic orientation of grains plays an important role. It is well known that the orientation of grains in dielectric and magnetic bodies affects their properties. When grains are aligned, the orientation is said to be high. When they are randomly oriented, orientation is low. When a Mn–Zn ferrite is hot pressed, the grains in the polycrystalline bodies exhibit a preferred orientation which is parallel to the pressing direction and along the $\langle 111 \rangle$ direction of the spinel structure as indicated in Fig. 2.63.

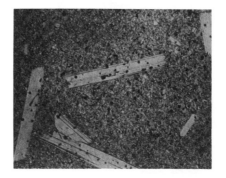

Fig. 2.62 Microstructure of β''-alumina. (From E. M. Vogel, D. W. Johnson Jr and M. F. Yan, *Am. Ceram. Soc. Bull.*, **60**, (4), 494, 1981. Reproduced by permission of The American Ceramic Society)

Fig. 2.63 Grain orientation of spinel ferrite. (From S. Hayakawa and Y. Matsuo, *Treatise of Ceramic Materials Technologies*, Industrial Technology Center, 1979, p. 218)

The preferred orientation of ferrites has been known to improve wear characteristics.

2.9 STRUCTURES OF POROUS BODIES

Porous bodies are those with a large number of pores. In contrast to dense polycrystalline bodies, porous bodies have been sought because of their unique functions which are listed in Table 2.15. Type 1 functions arise from the presence of fine pores with a large surface area, but it is not critical to control the size and size distribution of the pores. On the other hand, it is important to control the shape and size of the pores for Type 2 functions. Examples of porous bodies are listed in Table 2.16 together with ranges of pore sizes.

Table 2.15 Functions of porous ceramic bodies

Group 1 functions	Weight reduction		
	Thermal insulation	Lower-order functions	
	Sound absorption		Higher-order functions
	Vibration absorption		
	Shock absorption		
	Chemical absorption		
Group 2 functions	Selective chemical absorption		
	Ion exchange		
	Selective filtration		
	Selective transmission		
	Decomposition		

Table 2.16   Porous ceramic materials

Powders with fine pores		Sintered bodies	
Porous materials	Pore sizes (Å)	Porous materials	Pore sizes (μm)
Zeolite	3–15	Sintered alumina	200 Å–12
Silica gel	15–200	Sintered diatom earth	1–8
Alumina gel	40–400	Glass filters	5–200
Activated carbon	10–80	Sintered aluminosilicates	14–500
Silica-magnesia catalyst	10–100	Sintered carbon	15–170
Silica-alumina catalyst	70–250	Sintered silicates	100–600
Diatom earth	3000–10 μm		

Glasses		Fibers	
Porous materials	Pore sizes (Å)	Porous materials	Pore sizes (Å)
Porous glasses	15–2500	Glass fibers	1000–
Porous glass bodies for secondary electron amplification	–μm	Potassium titanate fibers	3–

In addition to crystallographic channels, pores are formed by sintering (open porosity) and by chemical treatment of phase-separated glasses.

2.9.1   Catalyst Carriers

Ceramic porous bodies with large surface areas have been increasingly utilized to improve the activity, selectivity and thermal endurance of catalysts. Examples of commercially important ceramic porous bodies are SiO_2, Al_2O_3, SiO_2–Al_2O_3 (aluminosilicates), zeolites, activated charcoal and MgO. In general, these porous bodies have a wide distribution of pore sizes. Zeolites have unusually narrow distributions of pore sizes because of the presence of crystallographic channels as indicated in Fig. 2.64.

One of the most important factors which control the activity and selectivity of a catalyst is the distribution of the catalyst on a carrier. Pore size affects the distribution of the catalyst as well as the absorption and desorption rates of gaseous molecules. Thus it is important to control the size and size distribution of the pores to optimize the catalytic functions.

2.9.2   Porous Glasses

When heat treated, borosilicate glasses separate into a phase of alkali borate and that of SiO_2. Porous bodies of SiO_2, which are called porous glasses, can be

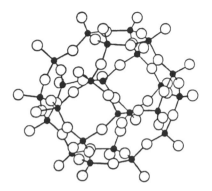

Fig. 2.64 Structure of a prototypic zeolite

obtained by dissolving the alkali borate phase in an acid. By varying the heat treatment, it is possible to obtain porous glasses with a large surface area and pore sizes between 10 and 3000 Å.

Porous glass has been used as a molecular filter to separate H_2 from H_2–H_2S gas mixtures. Mean-free paths of H_2 and H_2S at 25°C and one atmosphere are 1230 Å and 430 Å, respectively. Thus it is possible to separate these two gases by the pore size controlled filter. Inorganic–organic composite glasses can be made after infiltrating pores of the porous glasses with an organic monomer and polymerizing *in situ*. These composites are a good use of porous glasses.

2.9.3 Lightweight materials

Lightweight materials are used mainly for building. These have been developed to reduce weight and consist of bubble concrete, lightweight structural materials, bubble glass materials, various balloons and fly ash. In addition, materials with good thermal insulation, high heat resistance and sound-dampening capabilities have been developed continuously.

The sizes of the bubbles range from a few Ångstroms to a few millimeters and their morphologies also vary widely. An electron micrograph of a carbon balloon is shown in Fig. 2.65.

2.10 SURFACE STRUCTURES

Because a regular atomic arrangement is interrupted at a solid surface, it is necessary to treat the solid surface differently from the bulk structure. A solid surface can be viewed as a two-dimensional aggregate of lattice defects, but there are a number of aspects which need to be investigated in depth in the future.

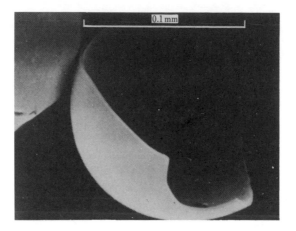

Fig. 2.65 Scanning electron micrograph of a fractured carbon balloon with a diameter of 0.17 mm. (Fine particles from the fracture can be seen in the balloon.) (From S. Koie, In *Treatise of Ceramic Materials Technologies*, Industrial Technology Center, 1979, p. 815)

Solid surfaces are of two types; as-grown free surfaces and treated surfaces. Very little roughness can be observed on clean surfaces such as the cleavage planes of single crystals. At an atomistic level the surfaces of ionic crystals consist of both anions and cations. Anions are squeezed out from the surfaces of LiF and MgO. As a result, a low-energy electron diffraction (LEED) study indicates that the bulk symmetry of the crystals is distorted. ZnO has an anisotropy along the c-axis and its c-planes are covered with either Zn or O ions exclusively. It is not possible to distinguish Zn or O planes by the X-ray diffraction method on the surfaces of polycrystalline ZnO with a preferred orientation of grains whose c-axis is perpendicular to the surface. However, they can be distinguished by etching the polycrystalline surface with an acid because of the difference in etching speeds between Zn and O planes. Thus it is very important to quantify the crystallographic orientation of polycrystalline surfaces when surface structures are discussed.

The structures of the surfaces which are polished, cut, or chemically treated are different from those of the free surfaces. For example, polishing creates surface layers (up to 20–30 Å deep), which are either amorphous or recrystallized with very fine grains. These are called Beilby layers and can be ascertained by the presence of hollows in electron diffraction patterns (see Fig. 2.66). When the surfaces are mechanically treated, they react with mechanical tools due to contact friction between them. As a result, surface compositions may change or new compounds may be formed. This is called the mechanochemical effect and makes the surfaces drastically different from the bulks.

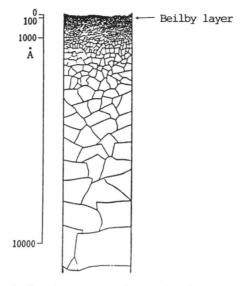

Fig. 2.66 Microstructural disturbance near the surface by polishing. (From N. Watanabe, A. Watanabe and Y. Tamai, *Surfaces and Interfaces*, Kyoritsu Publishing, 1973, p. 104: reproduced by permission of Kyroritsu Shuppan)

Solid surfaces are more or less reactive and react quickly with reactive gaseous species, such as moisture and oxygen, in the atmosphere. Thus clean surfaces cannot be maintained in air. For example, the surfaces of metal oxides are covered with OH^- ions. There are many instances where the surface characteristics are controlled exclusively by these surface-active elements which are quite different from the bulks, both compositionally and structurally. The importance of the surface-active elements extend one to several molecular layers deep and their amounts are usually less than one per cent of the total solid weights. It requires a small amount of gas to activate gas sensors, whose operations depend on resistance changes due to absorption–desorption of gases.

Silica, SiO_2, is a commercially important surface-active material and has two types of surface-active group, namely silanol and siloxane. As indicated in Fig. 2.67, there are three types of silanol bond. γ-Al_2O_3 is an important material as a catalyst as well as a catalyst carrier. IR spectra indicate five absorption bands due to the stretching–contractions of OH bonds as indicated in Fig. 2.68, whose model is summarized in Table 2.17.

Solid surfaces tend to attract impurities as well as lattice defects similar to grain boundaries. Their species and amounts depend on the sign and magnitude of surface charges. The structures of lattice defects at solid surfaces

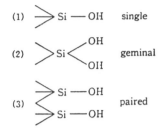

Fig. 2.67   Three types of silanol

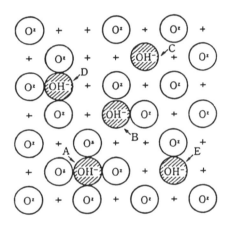

Fig. 2.68   Orientation model of OH⁻ ions (+ indicates Al^{3+} in the lower layer). (From J. B. Peri, *J. Phys. Chem.*, **69**, 211, 1965: ©1965 American Chemical Society)

Table 2.17   Coordination of OH⁻ on the surface of γ-Al_2O_3

Wave number (cm^{-1})	Coordination indicated in Fig. 2.68	Number of nearest-neighbor oxygen ions	Wave number (cm^{-1})	Coordination indicated in Fig. 2.68	Number of nearest-neighbor oxygen ions
3800	A	4	3780	D	3
3744	B	2	3733	E	1
3700	C	0			

J. B. Peri, *J. Phys. Chem.*, **69**, 211–20 (1965). ©1965 American Chemical Society.

have not been investigated extensively and is one of the topics requiring further research in the future.

2.11 THIN-FILM STRUCTURES

Thin films are important materials because of requirements for integration and miniaturization of electronic devices and are increasingly sought in electronic and optical applications. These films have essentially two-dimensional structures which are a few Ångstroms to a few micrometers thick. As a result, the surface effects influence the bulk properties significantly. Thus the properties of thin films are quite different from those of single crystals and polycrystalline bodies.

From an atomistic viewpoint the structures of thin films are single crystal, polycrystalline or amorphous. Depending on processing conditions, it is possible to obtain all three types of thin-film structure. Amorphous thin films are especially important as protective films and polycrystalline thin films with a high degree of orientation are useful as magnetic and dielectric materials.

Single-crystal thin films are grown on single-crystal substrates, such as NaCl, KCl, MgO, LiF and mica. These films may have some crystallographic orientations with single-crystal substrates. the presence of crystallographic orientations is called epitaxy, and this is observed when the thin films are grown above a critical temperature, which is called the epitaxy temperature. For example, when Ag is evaporated on the (110) cleavage plane of ZnO, the (110) plane of Ag grows parallel to the cleavage plane. The [001] direction of ZnS is also parallel to the [001] direction of Ag. These relationships can be summarized as follows:

$$(110)_{ZnS}//(110)_{Ag}$$

$$[001]_{ZnS}//[001]_{Ag}$$

There is a number of theories which have attempted to describe the mechanism of epitaxy. One is based on what is called the 'misfit', which is defined by the ratio of the lattice constant of a single crystal substrate, a, to the difference of the lattice constants between the substrate and the growing crystal, b. Thus the misfit coefficient is defined by

$$100x(b-a)/a \ (\%)$$

The Ag thin film on ZnS has an atomic arrangement as shown in Fig. 2.69 and its misfit coefficient is -25%. Although it was thought that the smaller the misfit, the more likely the epitaxial growth of thin films, it has been reported in a number of systems that it is possible to grow epitaxial thin films even if the misfit is large. Thus the misfit is not the only factor which controls the epitaxial

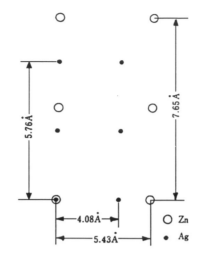

Fig. 2.69 Atomic arrangement of an Ag thin film on ZnS

growth of thin films. It has been reported that the large misfit can be accommodated by misfit dislocations, but this hypothesis has not been proved. Table 2.18 lists a number of examples of the epitaxial growth of metallic thin films evaporated on alkali halide substrates as a function of substrate temperature.

2.12 STRUCTURES OF INTERFACES

Many interesting materials have been made from interfacial structures. These structures vary widely and thus it might be confusing to discuss them without a proper classification. In this book, interfaces are classified as follows:

(1) Whether they are homogeneous or heterogeneous
(2) Whether they are open or closed
(3) Whether transport phenomena at the interfaces occur perpendicular or parallel to the interfaces.

This classification is summarized in Table 2.19.

Grain boundaries and twin boundaries are two examples of homogeneous closed interfaces. A grain-to-grain contact in a porous body is an example of homogeneous open interfaces. It is possible to transport chemicals to or away from the interface through open porosity. The transport of chemical substances in turn may cause the change in interfacial transport phenomena. A *p–n*

Table 2.18 Metallic thin films deposited on cleavage planes of alkali halides

Metal	Substrate	Preferred orientation			
		0	100	200	300
	KCl		– – –(111)$_s$,	(001)$_w$——(001)——	
Au	KBr		– – –(111)$_s$,	(001)$_w$——(001)——	
	KI		– – –(111)$_s$,	(001)$_w$——(001)——	
	KCl		————(001)————		
Ag	KBr		————(001)————		
	KI		————(001)————		
	KCl		– – – –(001)– – –+(001)——		
Cu	KBr		– – – –(001)– – –+(001)——		
	KI		– – – – (001)– – –+(001)——		
	KCl	– –(001)– – – –+		————(001)————	
Pd	KBr	– –(001)– – – –+		————(001)————	
	KI	– – – – – – – (001)$_s$– – –(111)$_w$– – – – –+(001)			
	KCl		– – – – – –(001)– – – – –+(001)		
Ni	KBr		– – – – – –(001)– – – – –+(001)		
	KI		– – – – – – –(001)– – – – – –		
	KCl		= = = (111)=+ (111), (001)		
Al	KBr		= = = = = =(111)= = = =+(111), (001)		
	KI		= = = = = = (111) = = = = =		

+– – –(hkl)– – –:(hkl) pattern mixed with Debye rings.
= = = (hkl)= = : almost-perfect fibrous structure.
————(hkl)————: (jkl) single-crystal structure.
From T. Kato, *Japanese Journal of Applied Physics*, 7, 1162, 1968. Reproduced by permission of *The Japanese Journal of Applied Physics*.

Table 2.19 Classification of interfaces

Classification	Direction of phenomena	
	Perpendicular to interface	Parallel to interface
Homogeneous		
Closed	GB⊥	GB//
Open	NK⊥	NK//
Heterogeneous		
Closed	HJ⊥	HJ//
Open	HC⊥	HC//

junction is an example of an heterogeneous closed interface. Although in many instances only electrical conduction from *p*- to *n*-type semiconductors is the subject of discussion, there are a number of interesting transport phenomena along the *p–n* junctions. For example, it is possible to alter some *p–n* transport phenomena such as electrical conduction by the transport of chemical substances to or from *p–n* junctions. The changes in the transport phenomena occur either perpendicular or parallel to p–n junctions. The classification of interfaces is shown schematically in Fig. 2.70. In the following some examples will be discussed.

2.12.1 NEW PHENOMENA BY HOMOGENEOUS CLOSED INTERFACES

(1) Those phenomena which occur perpendicular to the interface (corresponding to GB$\perp$ in Fig. 2.70(a)): SiC has been investigated extensively as a structural material at high temperatures. Researchers at the Hitachi Research Laboratory have found that it is possible to make polycrystalline SiC with high thermal conductivity by hot pressing SiC with a small amount of BeO as a sintering additive. Polycrystalline SiC has been investigated as a material for making high thermal conductivity IC substrates.

The SiC grains in the polycrystalline SiC exhibit semiconductivity, but the grain boundaries are electrically insulating. While the polycrystalline SiC sintered without BeO is an n-type semiconductor, the polycrystalline SiC sintered with BeO is p-type. The lattice structure of SiC grains is without any defects up to grain boundaries. The presence of any BeO phases cannot be detected along the grain boundaries. The solid solubility of BeO in SiC has not been observed and it has been demonstrated that BeO exists in the grain boundaries. Crystallographically it is anticipated that pure SiC might have high thermal conductivity, but because of the dissolution of a small amount of nitrogen and low sintered density, usual SiC does not have high thermal conductivity. On the other hand, BeO functions effectively to remove nitrogen from the SiC grain matrix and to promote sintering as a good sintering aid. SiC lattices are without any defects to the grain boundaries and thus the movement of phonons for thermal conduction is not affected significantly by the grain boundaries. On the other hand, the movement of electrons for electrical conduction is hindered by the presence of p-type regions or barriers.

Researchers at the Matsushita Electric Wireless Research Laboratory invented ZnO varistors. These varistors are sintered with small amounts of bismuth and antimony oxides, have very non-linear voltage–current characteristics and have been used as overvoltage protection circuit elements. It is expected that ZnO varistors have structure similar to the polycrystalline SiC with BeO.

Semiconducting polycrystalline barium titanate has low electrical resistivity at temperatures below the Curie temperature but high electrical resistivity above it. Thus intelligent elements which function as a heater, a temperature sensor and an actuator can be manufactured from polycrystalline barium titanate. The elements have been used widely in

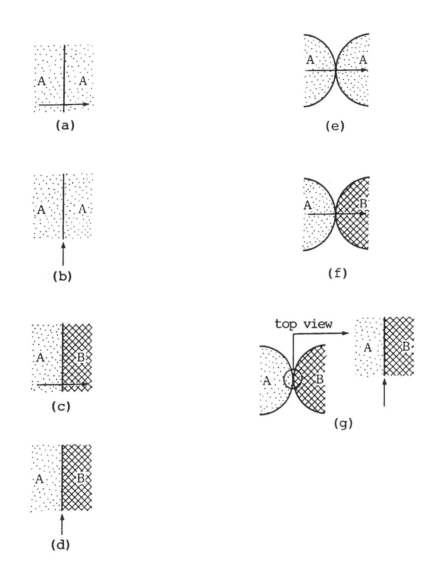

Fig. 2.70 Schematics of various interfacial structures. Shown are phenomena which occur (a) perpendicular to a homogeneous closed interface (GB⊥); (b) parallel to a homogeneous closed interface (GB//); (c) perpendicular to a heterogeneous closed interface (HJ⊥); (d) parallel to a heterogeneous closed interface (HJ//); (e) perpendicular to a homogeneous open interface (HK⊥); (f) perpendicular to a heterogeneous open interface (HC⊥); (g) parallel to a heterogeneous open interface (HC//). (From H. Yanagida, *Chemistry and Industry*, **39**(3), 831, 1986)

hair and futon dryers. This so-called PTC effect is also due to the formation of electrical barriers along grain boundaries.

(2) Those phenomena which occur parallel to interfaces (corresponding to GB// in Fig. 2.70(b): In most polycrystalline systems grain boundary diffusion is faster than grain matrix diffusion. During sintering of ZnO varistors and polycrystalline barium titanates, extreme care has been exercised to allow the diffusion of oxide ions along grain boundaries so that insulating barriers can be formed. This phenomenon is important in processing.

2.12.2 New Phenomena by Heterogeneous Closed Interfaces

(1) Those phenomena which occur perpendicular to interfaces (corresponding to HJ⊥ in Fig. 2.70(c)): Heterojunctions belong to this class. Unlike heterojunctions at open interfaces, which will be discussed later, CuO–ZnO interfaces exhibit a normal heterojunction behavior with typical current–voltage characteristics. Because of the closed interfaces these phenomena are not affected by atmosphere.

(2) Those phenomena which occur parallel to interfaces (corresponding to HJ// in Fig. 2.70(d): Grain boundary diffusion between two dissimilar substances is also enhanced, which is analogous to homogeneous grain boundary diffusion. Ionic conductivity can also be improved by dispersing insulators in ionic conductors, which is due to the formation of ionic conductor–insulator interfaces.

2.12.3 New Phenomena by Homogeneous Open Interfaces

Those phenomena which occur perpendicular to interfaces (corresponding to HK⊥ in Fig. 2.70(e)) will be discussed in this section. The combustible gas sensors made of semiconducting ZnO exhibit their sensor characteristics because of the sensitivity of grain-to-grain contacts (neck areas) to combustible gases in the atmosphere. The ZnO grains exhibit n-type semiconducting behavior and the neck areas have high electrical resistance due to the formation of n–p–n or n–i–n junctions caused by the adsorption of oxygen in air. When combustible gases are present in the atmosphere, these gases promote the desorption of oxygen from neck areas, which causes the elimination of either p or i and thus decreases electrical resistance. Sensor functions have been optimized by controlling porosity and electrical conductivity, which can be accomplished by two-step doping of Al_2O_3 and Li_2O. This processing approach has been based on the interfacial mechanism of the sensors.

2.12.4   New Phenomena by Heterogeneous Open Interfaces

(1)   Those phenomena which occur perpendicular to interfaces (corresponding to HC⊥ in Fig. 2.70(f)): Yanagida *et al.* discovered that moisture significantly influences the voltage–current characteristics of open heterojunctions of the CuO–ZnO system (Fig. 2.71). This phenomenon is best described as follows. Adsorbed water molecules receive electron holes from the *p*-type semiconductor and form protons, which migrate to the *n*-type semiconductor and recombine with electrons to form hydrogen molecules. (Oxygen molecules are generated at the *p*-type semiconductor.) Since adsorbed water molecules are continuously electrolyzed, it is not necessary to reheat the sensors repeatedly to remove adsorbed moisture. These open heterojunctions are also effective in detecting the presence of CO_2.

(2)   Those phenomena which occur parallel to interfaces (corresponding to HC// in Fig. 2.70(g)): Moisture sensors made of a mixture of an acidic refractory, TiO_2, and a basic refractory, $MgCr_2O_4$, have been reported to be superior in long-term reliability and sensitivity and have been used in microwave ovens. Adsorbed water molecules are said to be dissociated at the TiO_2–$MgCr_2O_4$ interfaces and thus electrical conductivity increases along the interfaces.

2.13   STRUCTURES OF FIBERS

Structures of fibers are either single crystal, polycrystalline (with and without preferred orientation) or amorphous. Some fibers can be made infinitely long and others are very short. Glass fibers are the most widely known ceramic fibers and have been used extensively to make reinforced composites and thermal insulators. They are also useful for reducing weight and saving energy. Optical fibers, which have made remarkable progress during the last decade, are long fibers of silica glass (SiO_2). Optical fibers have two types of structure as shown in Fig. 2.72. In Fig. 2.72(a) the optical fiber is made of a core with a high refractive index and a cladding which surrounds the core and has a low refractive index. Light incident to one end of the optical fiber will experience total reflection repeatedly as it travels through the fiber. On the other hand, light incident to the fiber with a graded structure indicated in Fig. 2.72(b) travels without any delay.

Carbon fibers are another familiar group and have been used widely to make shafts for golf clubs. These have three types of structure as shown in Fig. 2.73 and are fibers whose structures are very close to those of graphite and polycrystalline fibers with or without a preferred orientation. Elastic moduli of

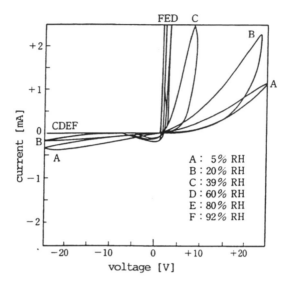

Fig. 2.71 Effects of moisture on the voltage–current characteristics of CuO–ZnO heterojunctions. (From Y. Nakamura, A. Ikejiri, M. Miyayama, K. Koumoto and H. Yanagida, *J. Jap. Chem. Soc.*, **1985**(6), 1154)

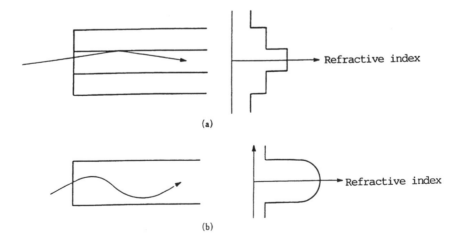

Fig. 2.72 Structures of optical fibers

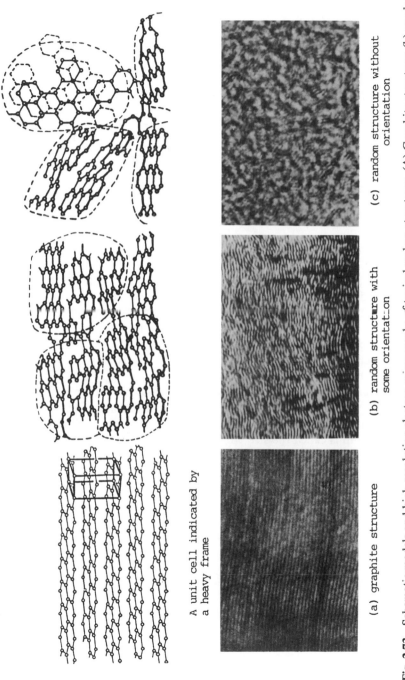

A unit cell indicated by
a heavy frame

(a) graphite structure

(b) random structure with
some orientation

(c) random structure without
orientation

Fig. 2.73 Schematic models and high-resolution electron micrographs of typical carbon structures. (A) Graphite structure; (b) random structure with some orientation; (c) random structure without orientation. (Courtesy of Professor Emeritus S. Ohtani of Gunma University)

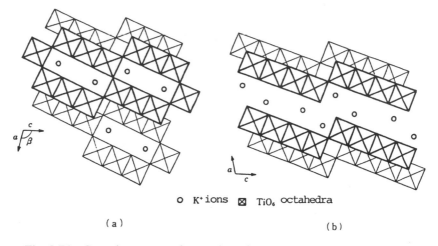

○ K$^+$ ions ⊠ TiO$_6$ octahedra

(a)   (b)

Fig. 2.74   Crystal structure of potassium titanate. (a) Kt$_2$Ti$_6$O$_{13}$; (b) K$_2$Ti$_4$O$_9$

the fibers depend on fiber structure and various carbon fibers have been manufactured for a variety of applications.

Single-crystal fibers can be classified into two types, namely those fibers whose crystal structures are fibrous and those produced by special techniques. An example of the former is potassium titanate, K$_2$O $\cdot$ nTiO$_2$. Their structures have either channels or layered spaces along the b-axis (Fig. 2.74). Alumina fibers are an example of the latter.

Whiskers are not only single crystals but must also have a screw dislocation along the growth direction. Whiskers of Al$_2$O$_3$ and SiC are widely known. In general, the strengths of whiskers are inversely proportional to their diameters. This relationship is called the size effect. It is also known that the strengths of whiskers depend on crystallographic orientation. At present it is not possible to mass produce whiskers. Furthermore, other, superior, fibers have been made available. As a result, whiskers have not been used commercially.

3

REACTIONS IN CERAMICS

3.1 PHASE EQUILIBRIA AND PHASE DIAGRAMS

A 'system' is defined as an assembly of substances that are isolated and independent from a 'surrounding'. Systems can be classified as either homogeneous or heterogeneous. A heterogeneous state consists of homogeneous states. The homogeneous state of a system is called the phase. Thus a system consists of either a single phase or multiple phases. Substances in a system are called the components and their mass ratios are called either the composition or the concentration.

Depending on the number of components, a system is called single-component, two-component, three-component, etc. Components are made of substances and do not necessarily imply atoms. The composition (or the concentration) of a system can be varied independently. For example, consider an aqueous solution of hydrochloric acid. This system consists of two components, namely hydrochloric acid (HCl) and water. However, the system cannot be treated as a three-component system made up of H, Cl and O simply because the ratios among H, Cl and O in an aqueous solution cannot be varied independently.

3.1.1 Phase Rule

When the free energy of a system is at its minimum, the system is in equilibrium. The free energy of a system G can be given by

$$G = H - TS \tag{3.1}$$

where H, S and T are the enthalpy, entropy and absolute temperature, respectively.

J. W. Gibbs derived a relationship between a number of components C and a number of phases P in an equilibrium state, given by

$$F = C - P + 2 \qquad (3.2)$$

where F is the degree of freedom. This relationship is thus called the Gibbs phase rule. The degree of freedom can be defined as the number of variables which can be varied independently without changing the number of phases in a system. Independent variables are temperature, pressure and composition.

For a system made of solid NaCl, both component and phase are one. Thus

$$F = 1 - 1 + 2 = 2$$

Thus it is necessary to fix two independent variables to define the system uniquely: it is possible to vary both temperature and pressure independently in the system. On the other hand, when a solid phase and a liquid phase co-exist, $C = 1$ and $P = 2$. Hence,

$$F = 1 - 2 + 2 = 1$$

which implies that the temperature of the system is set once the pressure is set or the pressure of the system is set once the temperature is set. If the two variables are varied independently, the system will return to either a solid or a liquid phase. It is quite natural, from the phase rule, that there is only one melting point for a solid at one atmosphere.

3.1.2 One-component Phase Diagrams

The phase diagram for H_2O is shown in Fig. 3.1. According to the phase rule, the degree of freedom is two in the regions where only one phase, ice, water or water vapor, is present and thus it is possible to change both temperature and pressure independently without affecting the stability of the phases. The degree of freedom is one where a combination of phases, either water/ice, water/water

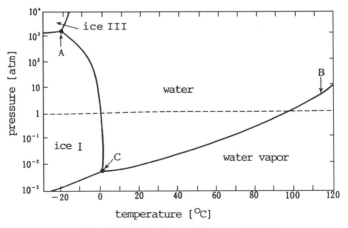

Fig. 3.1 The phase diagram of H_2O

vapor or ice/water vapor, is present and thus it is possible to change only a single parameter, either temperature or pressure, independently. In the figure the curves AC and BC correspond to these two-phase lines. When three phases, ice, water and water vapor, are present, the degree of freedom is zero and thus both temperature and pressure are fixed. In the figure point C corresponds to this condition and is called the triple point. It is well known that the triple point of water is at 0.0075°C and 4.58 mmHg.

3.1.3 Two-component Phase Diagrams

Technologically, solid and liquid states are two important phases of materials. In addition, physical and chemical properties are discussed for materials at atmospheric pressure. Hence in many cases it is possible to disregard the presence of a gaseous phase. In this section, two-component phase diagrams will be discussed at a total pressure of one atmosphere. Where the presence of a gaseous phase is not important, the phase rule for two-component systems becomes

$$F = 3 - P$$

It is customary in two-component phase diagrams to plot compositions along the abscissa and temperatures along the ordinate. In the following several important phase diagrams will be discussed for two-component systems.

3.1.3.1 COMPLETE SOLID SOLUTION DIAGRAMS

A phase diagram of systems which show complete solid solubility is given in Fig. 3.2. This phase diagram is applicable to systems wihch form solid solutions regardless of the compositions between substances A and B. Points P and Q in the figure indicate the melting points of pure A and B, respectively.

Let us consider the slow cooling of a liquid with an intermediate composition between A and B in Fig. 3.3. When a liquid with a composition x is cooled from T_1 to T_2, a solid phase with a composition y starts to precipitate. When the system is cooled further to a temperature T_3, a liquid phase with a composition u and a solid phase with a composition w co-exist. The ratio between the liquid and solid phases at the temperature can be given by

$$\text{Amount of liquid/amount of solid} = de/cd$$

This expression is called the lever rule. When the system is cooled to a temperature T_4, the system has only the solid solution with a composition x. The upper curve which connects P and Q in Fig. 3.3 is called the liquidus and the lower curve the solidus. Two phases co-exist in the region which is delineated by these two curves. Thus $F = 3 - P = 1$, which indicates that once the temperature is specified, the composition is automatically fixed.

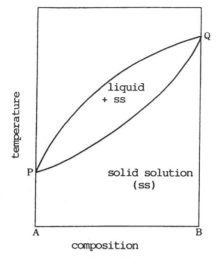

Fig. 3.2   Two component phase diagram with complete solid solubility

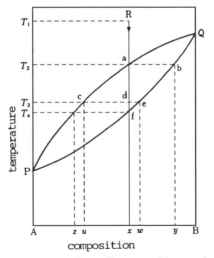

Fig. 3.3   Two-component phase diagram with complete solid solubility

This type of phase diagram can be seen in systems made up of two components with the same crystal structure and close ionic radius such as MgO–NiO and NaCl–KCl.

3.1.3.2   EUTECTIC DIAGRAMS

The phase diagram shown in Fig. 3.4 is called the eutectic type. Let us consider the cooling of a liquid with a composition x which has the lowest liquidus

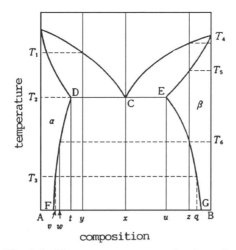

Fig. 3.4 Two component eutectic phase diagram

temperature indicated by point C in the figure. At T_2 a solid solution α with a composition t and a solid solution β with a composition u precipitate out simultaneously. Since $P = 3$, $F = 0$, which indicates that the temperature will be maintained constant until the precipitation is completed. When the system is cooled further, the system consists of two solid solutions, namely solid solution α with a composition which changes along curve DF and solid solution β with a composition which changes along curve EG. Curves DF and EG are called the solid solubility limits. At T_3 the composition of α is given by v and the composition of β by q. The reaction at point C is expressed as

$$\text{Liquid (C)} = \text{solid (D) (ss } \alpha) + \text{solid (E) (ss } \beta)$$

where ss denotes a solid solution. Since two solid solutions α and β precipitate simultaneously, the reaction is called eutectic. The resulting solids are called eutectic and point C is the eutectic temperature.

When a liquid with a composition z is cooled, the solid solution β starts to precipitate at T_4 and the amount of the liquid continues decreasing until it reaches T_5. Between T_5 and T_6 only a solid solution with a composition z exists. At T_6 the solid solution α with a composition w starts to precipitate. The compositions and amounts of the two solid solutions change when the system is cooled further. This type of phase diagram can be seen in the system between CaO and MgO.

3.1.3.3 PERITECTIC DIAGRAMS

The phase diagram shown in Fig. 3.5 is called the peritectic type. Let us consider the cooling of a liquid with a composition x which passes point P in

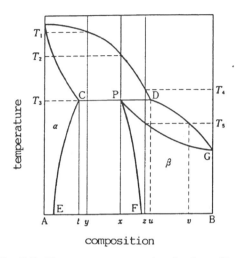

Fig. 3.5 Two component peritectic phase diagram

the figure. At T_2 a solid solution α starts to precipitate. As it cools further, the composition and amount of the solid solution change along the solidus and liquidus of the system. At T_3 a solid solution β precipitates along the outer surfaces of the α precipitates until the system is completely converted to the β precipitates at the temperature. When the system is cooled further, the solid solution α reprecipitates out of the β precipitates. The composition and amount of the precipitates changes along the solid solubility limit curves. (The reaction which represents the precipitation of a solid phase from another solid phase during the cooling is called the dissolution–precipitation reaction.) The reaction at point P can be given by

$$\text{Liquid (D)} + \text{solid (C) (ss } \alpha\text{)} = \text{solid (P) (ss } \beta\text{)}$$

Since β precipitates as if to surround α, this is called the peritectic reaction. Point P is called the peritectic point and T_3 the peritectic temperature.

When a liquid with a composition z is cooled from a high temperature, the precipitation of α starts at T_4. The peritectic reaction occurs at T_3. The system remains at the temperature until all α precipitates are converted to β precipitates with a composition x, which co-exists with a liquid phase with a composition u. As it cools further, the compositions and amounts of the β precipitates and the liquid phase change along the liquidus and solidus curves. At T_5 the system has only the β phase.

Peritectic phase diagrams are seen in systems such as AgCl–LiCl, Bi_2O_3–La_2O_3, and ZrO_2–La_2O_3.

3.1.3.4   MONOTECTIC DIAGRAMS

A phase diagram with a monotectic reaction is shown in Fig. 3.6. Let us consider the cooling of a composition x from a high temperature. At T_1 the liquid phase separates into two liquid phases. At T_2 the liquid phases have the compositions of M and D. At this temperature the liquid phase M undergoes the following reaction:

$$\text{Liquid (M)} = \text{solid (C) (ss } \alpha) + \text{liquid (D)}$$

This is similar to the peritectic reaction, but because of the formation of a solid and a liquid, it is called the monotectic reaction. Point M is called the monotectic point and T_2 is the monotectic temperature. Further cooling after completion of the monotectic reaction will cause precipitation of solid solutions α and β from the liquid phase F at T_3. Since some α precipitates exist prior to the coprecipitation of α and β, solid solutions α and β precipitate around the α precipitates which are formed above the monotectic temperature.

This type of phase diagram has been observed in systems between an alkali metal and its halide such as K–KBr and Na–NaCl and oxide systems such as CaO–SiO_2, ZrO_2–SiO_2, and PbO–B_2O_3.

3.1.3.5   OTHER DIAGRAMS

In addition to the four basic reactions in two-component systems discussed above, there exist the following special reactions:

(1)   Metatectic: solid (I) = solid (II) + liquid
(2)   Eutectoid: solid (I) = solid (II) + solid (III)

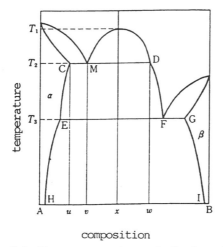

composition

Fig. 3.6   Two component monotectic phase diagram

(3)   Peritectoid: solid (I) + solid (II) = solid (III)

In reaction (1) a solid solution dissociates into another solid solution and a liquid at lower temperatures. Reactions (2) and (3) are similar to the eutectic and peritectic reactions, but unlike these, the reactions occur only in solids.

   Thus far, simplified two-component phase diagrams have been illustrated. Phase diagrams for real systems are usually more complex and involve a combination of reactions. By applying the phase and the lever rules judiciously, these phase diagrams can be understood easily. The phase diagram of the BaO–TiO₂ system is shown in Fig. 3.7.

3.1.4 Three-component Phase Diagrams

In three-component systems $C = 3$ and thus the degree of freedom is given by $F = 4 - P$. The compositions of three-component systems can be illustrated schematically by points in a right triangle as shown in Fig. 3.8. The composition at point P in the figure is A 25%, B 50% and C 25%. The ratios between two components on a straight line which passes through an apex

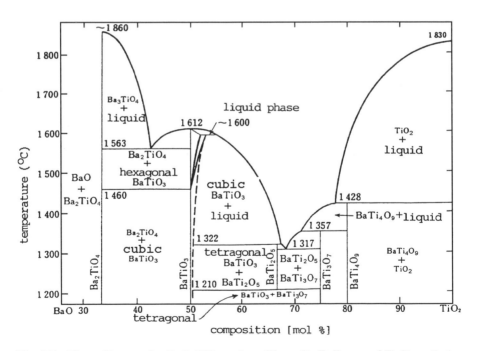

Fig. 3.7   Phase diagram for BaO–TiO₂ system. (From D. E. Rase and R. Roy, *J. Am. Ceram. Soc.*, **38**(3), 111, 1955. Reproduced by permission of The American Ceramic Society)

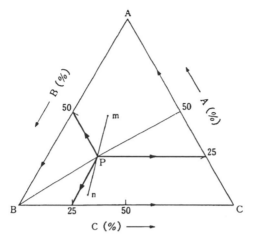

Fig. 3.8 Reading of a composition in a three-component phase diagram

of the triangle are constant and the composition of a component is constant on a line parallel to an edge of a triangle. The lever rule can be applied also to three-component phase diagrams. Assuming that a solid at point P consists of two phases, namely m and n, the ratio of these two phases can be given by

$$(\text{Phase m})/(\text{phase n}) = np/mp$$

In three-component phase diagrams the temperature axis is perpendicular to the plane of the triangle and thus the phase diagrams are three-dimensional. Let us consider a phase diagram of systems which exhibit complete solid solubility (Fig. 3.9). In the figure the upper plane delineated by A'B'C' is called the liquidus plane and the lower plane is the solidus plane. When a liquid with a composition x is cooled slowly from a high temperature, solid a' starts to precipitate at the temperature corresponding to point a. Point a' is located on an intersection, c'd', between an isothermal plane and a solidus plane for point a, but its exact composition cannot be known *a priori* and has to be determined experimentally because the system has two degrees of freedom. When the system is cooled further, the liquid composition changes from a to b and the solid composition from a' to b'. At the temperature for point b' the liquid phase is completely converted to a solid phase which has the composition of x. The solid phase maintains this composition upon further cooling.

Next, consider a system which exhibits complete liquid solubility but does not dissolve each other at all in solid phases. The phase diagram which exhibits these characteristics is called the eutectic type and is shown in Fig. 3.10. When a liquid with a composition x is cooled from a high temperature, component C starts to precipitate at the temperature corresponding to point a. Since only C precipitates, the ratio between a and b in the liquid does not change during

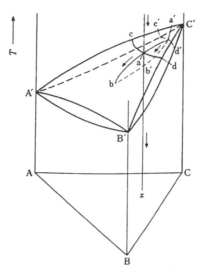

Fig. 3.9 Three-component phase diagram with complete solid solubility

cooling. Thus the liquid composition changes along the intersection ab which is formed by a plane CC'a and a liquidus plane $C'E_3'E'E_2'$. At the temperature corresponding to point b both A and C precipitate around the C precipitates, which are already present in the liquid. The liquid composition changes along bE' upon further cooling. At point E', B precipitates. The system is completely solidified into a mixture of A, B and C at this temperature.

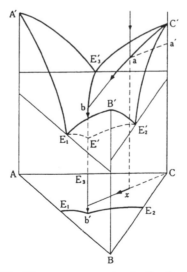

Fig. 3.10 Three-component eutectic phase diagram

Phase diagrams for three-component systems are usually quite complicated and it becomes very difficult to show them three-dimensionally. In order to exhibit three-component phase diagrams two-dimensionally, the phase diagrams are often shown at various isothermal planes. In some cases the phase diagrams are shown as pseudo-binary phase diagrams by making one of the three components constant. In other cases isothermal planes are projected onto the composition triangle. It is customary in these projections to show only the intersections with liquidus planes.

3.2 PHASE TRANSFORMATIONS

The change of a homogeneous substance from one phase to another at a certain temperature and pressure is called phase transformation. Most common phase transformations at atmospheric pressure are the change from ice to water at 0°C and from water to water vapor at 100°C. Although a number of phase transformations among gas, liquid and solid phases are possible, only the solid-to-solid phase transformations will be discussed in this section.

It is essential to look at phase transformations from both thermodynamic and crystallographic viewpoints.

3.2.1 Thermodynamic Views of Phase Transformations

Let us consider the phase transformations of a solid A from A_1 to A_2 to A_3 phases when heated from a low temperature. The phase transformation of a

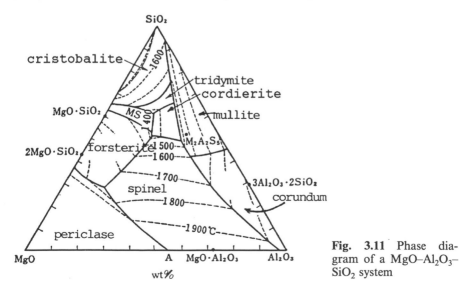

Fig. 3.11 Phase diagram of a MgO–Al$_2$O$_3$–SiO$_2$ system

solid which involves changes in crystal structures but without compositional changes is called polymorphic transformation. The magnitudes of free energy of various phases determine which is the most stable phase at a given condition.

The free energy of a solid at temperature, T, and pressure, P, can be given by

$$G = E + PV - TS \qquad (3.3)$$

where E is the internal energy, S the entropy and V the volume. Except those phase transformations at very high pressures, it is possible to ignore the contribution of the product PV simply because the term is usually much smaller than the other two. At absolute zero temperature the free energy is equal to the internal energy. With increasing temperature the contribution of the product TS to G becomes larger and thus structures with higher entropies tend to decrease G further than those with smaller entropies. Temperature dependencies of the internal energy and the free energy of A_1, A_2 and A_3 phases are shown schematically in Fig. 3.12. In the figure the entropy contribution is represented by $E - G$. It is evident that stable phases at high temperatures have crystal structures with a large internal energy and entropy.

The phase transformation from phase I to phase II occurs when the free energy of each phase are equal; namely,

$$G_I(T, P) = G_{II}(T, P)$$

where T and P are the temperature and pressure at which the phase transformation occurs, respectively. Thermodynamically, there are two classes of phase transformations. Phase transformations are called primary when they satisfy

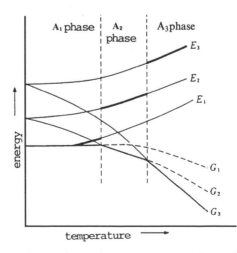

Fig. 3.12 Dependence of energies on temperature for various polymorphs

$$(\partial G_I/\partial T)_P \neq (\partial G_{II}/\partial T)_P \text{ and } (\partial G_I/\partial P)_T \neq (\partial G_{II}/\partial P)_T$$

and are called secondary when they satisfy

$$(\partial G_I/\partial T)_P = (\partial G_{II}/\partial T)_P \text{ and } (\partial^2 G_I/\partial T^2)_P \neq (\partial^2 G_{II}/\partial T^2)_P$$

Primary phase transformations have discontinuities in enthalpy $H = E + PV$ and volume V. On the other hand, secondary phase transformations have discontinuities in first derivatives of some thermodynamic functions such as specific heat dH/dT and thermal expansion dV/dT. The polymorphic phase transformations discussed earlier are primary. The formation of a glassy phase from a liquid is an example of secondary phase transformation (see Section 2.4).

The curve which represents the dependence of transformation temperature on pressure is called the transformation curve. The curves for primary phase transformations can be given by the following Clausius–Clapeyron equation:

$$dP/dT = \Delta H/T\Delta V \tag{3.4}$$

where ΔH is the heat of melting, or heat of evaporation, or heat of transformation. The transformation curves of SiO_2 are shown in Fig. 3.13. In the figure the transformation curves are shown by bold lines. Among the five phases of SiO_2 including a liquid, the phase which has the least vapor pressure at a given temperature is the stable phase at that temperature. The

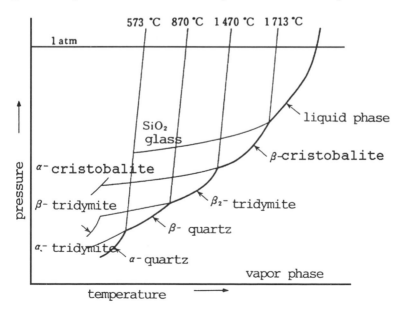

Fig. 3.13 Phase transformations of SiO_2

transformation temperatures among various phases at one atmosphere are also indicated in the figure. The transformation curves are for transformations under equilibrium and not for those under non-equilibrium conditions. Normally silica glass is formed when a liquid phase of SiO_2 is cooled from a high temperature. When cooled, β-cristobalite transforms to α-cristobalite and β_2-tridymite transforms to β- and α-tridymite. These phases are called metastable. In Fig. 3.13 the P-T curves for the metastable phases are shown schematically by fine lines.

A metastable phase transforms to a more stable one with less free energy at a given temperature and this transformation is governed by the kinetic processes involved. The mobilities of various atoms play a decisive role in determining the rate of phase transformation and thus it is very important to consider the changes in crystallographic structure, which will be the subject of the next section.

3.2.2 Crystallographic Views of Phase Transformations

The rate of phase transformation of a solid from one crystal structure to another depends strongly on the geometric relationship between them. Some phase transformations involve minor displacements of atomic positions. Others may involve breakage and reformation of chemical bonds. Thus the activation energy associated with phase transformation plays an important role in determining the rate of transformation.

Phase transformations are classified by comparing crystal structures among polymorphs as indicated in Table 3.1.

Table 3.1 Classification of phase transformations

Type of phase transformation	Examples	Rate of phase transformation
(i) PT involving secondary coordination structures		
(a) Displacive	ZrO_2 (monoclinic $\leftrightarrow$ tetragonal)	Fast
	$BaTiO_3$ (tetragonal $\leftrightarrow$ cubic)	
	α-quartz $\leftrightarrow$ β-quartz	
	β-cristobalite $\leftrightarrow$ α-cristobalite	
(b) Reconstructive	β-quartz $\leftrightarrow$ β_2-tridymite	Slow
	β_2-tridymite $\leftrightarrow$ β-cristobalite	
	TiO_2 (antase $\leftrightarrow$ rutile)	
(ii) PT involving primary coordination structures		
(a) Dilatational	CsCl (high temp. $\leftrightarrow$ low temp.)	Fast
(b) Reconstructive	$CaCO_3$ (aragonite $\leftrightarrow$ calcite)	Slow
(iii) Order–disorder PT	Cu_3Au, CuAu, etc.	Fast
(iv) PT involving the change in chemical bond type	C (diamond $\leftrightarrow$ graphite)	Slow

3.2.2.1 PHASE TRANSFORMATIONS WHICH INVOLVE CHANGES IN SECONDARY COORDINATION STRUCTURES

In this class of phase transformation the number and arrangement of nearest-neighbor atoms are not disturbed, but the atomic arrangements of atoms situated further away are altered. There are two types of phase transformation in this class: displacive or reconstructive. As indicated in Fig. 3.14, a displacive phase transformation involves minor distortion of structure (b) to structure (c). The range of atomic movement and the energy barrier for phase transformation are quite small in displacive phase transformations, thus these transformations occur very rapidly. Examples of this type of phase transformation are α to β transformation of quartz and cubic to tetragonal transformation of $BaTiO_3$.

Reconstructive phase transformation is illustrated schematically in Fig. 3.14 (a) to (b). In this type of phase transformation it is necessary to break chemical bonds and to rearrange the crystal structure in a primary coordination scale. Because of the necessity to overcome a very high activation energy, the rates of this type of phase transformation are quite low. Transformations among β-quartz, β_2-tridymite and β-cristobalite are examples of this type of phase transformation. The fact that β-cristobalite transforms to metastable α-cristobalite instead of stable β_2-tridymite upon cooling can be easily understood from a crystallographic viewpoint.

3.2.2.2 PHASE TRANSFORMATIONS WHICH INVOLVE CHANGES IN PRIMARY COORDINATION STRUCTURES

There are two types of phase transformation which involve the alteration of coordination number during transformation: dilatational or reconstructive.

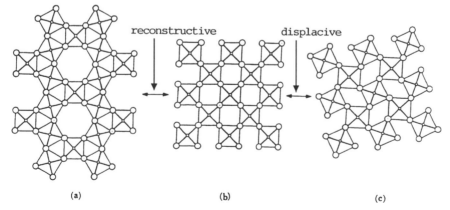

(a)　　　　　　　　　(b)　　　　　　　　　(c)

Fig. 3.14 Phase transformations of secondary coordination structures. (From T. Asamaki, *Basic Making of Thin Film*, Nikkan Kogyo Shinbunsha Shuppankyoku, 1979, p. 61: reproduced with permission)

The transformation of CsCl is a good example of dilatational phase transformation. As indicated in Fig. 3.15, Cs ions are coordinated with eight Cl ions in the low-temperature form of CsCl, which transforms to the high-temperature form with a NaCl structure at 460°C. Two Cl^- ions located at two diagonal corners of a cube are elongated along the threefold rotation axis and the remaining six ions are displaced at an equidistance around the Cs ion at the body center. Cs^+ ions are coordinated with eight Cl^- and Cl^- ions with eight Cs^+ ions in the low-temperature form. On the other hand, both ions are coordinated with six ions in the high-temperature form. The transformation of Fe from bcc to fcc is another example of this type of phase transformation. The rate of this type of phase transformation is high.

There is no simple mechanism in reconstructive transformation. During transformation chemical bonds between the nearest neighbors are broken and reformed one by one. Aragonite to calcite transformation of $CaCO_3$ is an example of this type of phase transformation and the rate of transformation is very low.

3.2.2.3 ORDER–DISORDER PHASE TRANSFORMATIONS

Cu and Au form a solid solution with complete solid solubility and both occupy the face centered cubic positions randomly in the high-temperature form of the alloy with a composition of Cu_3Au. When the alloy is cooled slowly, Au atoms occupy the corner positions of the face centered cubic lattice and Cu atoms occupy the face centered positions (Fig. 3.16). Thus this type of phase transformation is called order–disorder.

3.2.2.4 PHASE TRANSFORMATIONS WHICH INVOLVE A CHANGE IN CHEMICAL BOND

Metallic to non-metallic transformation is a good example of phase transformation which involves two polymorphs with different chemical

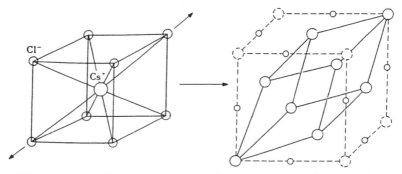

Fig. 3.15 Transformation of low- to high-temperature phases of CsCl

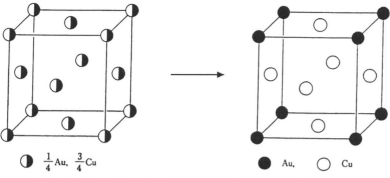

Fig. 3.16 Order–disorder phase transformation of 25Au–75Cu

bonds. The graphite-to-diamond transformation is a well-known example of this type of phase transformation. C atoms in graphite are joined together by chemical bonds formed by sp^2 orbitals and a van der Waals force. Diamond has chemical bonds formed by sp^3 orbitals. This type of phase transformation is very slow.

3.2.3 Nucleation

When a new phase appears from a homogeneous phase by changes in temperature and pressure, very small regions of the new phase are created by compositional fluctuation to start a phase transformation. When a new stable phase appears from an unstable or metastable host phase, it is expected to decrease in free energy $\Delta g_v (< 0)$ per unit volume as well as increase in free energy, σ, per unit area due to the formation of interfaces between the host and the new phases. Thus depending on the size of the fluctuation, the change in total free energy may or may not be negative. In addition, an increase in free energy due to strain has to be considered for phase transformations with changes in volume.

Let us consider the formation of a new phase from a host phase without taking into account the increase in free energy due to strain. When a spherical new phase with a radius r is formed in a host phase, the change in total free energy can be given by

$$\Delta G = 4\pi r^2 \sigma + (4/3)\pi r^3 \Delta g_v \qquad (3.5)$$

As indicated in Fig. 3.17, the total free energy increases with increasing r until it reaches a value r^*, but decreases beyond r^*. Here, r^* is called the critical radius for nucleation. Regions of the new phase having a radius less than r^* are called embryos: those of radius r^* and larger are called nuclei. From $d\Delta G / dr = 0$,

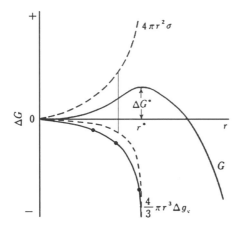

Fig. 3.17 Free energy of a spherical nucleus

$$r^* = -2\sigma/\Delta g_v \tag{3.6}$$

$$\Delta G^* = 16\pi\sigma^3/3(\Delta g_v)^2 \tag{3.7}$$

where $\Delta g_v = \Delta h_v - T\Delta S_v$. Since Δh_v and ΔS_v exhibit little temperature dependence within a limited range of temperature, the absolute value of Δg_v increases with decreasing temperature. Furthermore, no significant temperature dependence of σ is expected. Thus from Eqs (3.6) and (3.7), both ΔG^* and r^* become smaller with increasing supercooling.

The number of the critical nuclei of radius r^* is proportional to the Boltzmann factor $\exp(-\Delta G^*/kT)$. The critical nuclei grow with further migration of atoms from the host phase and become stable. The rate of atomic migration depends also on the Boltzmann factor $\exp(-\Delta G_m/kT)$, where ΔG_m is the activation free energy for the diffusion and migration of atoms. Thus the number of nuclei formed in a host phase in a unit volume per unit time (the nucleation rate) N can be given by

$$N \propto \exp\{-(\Delta G^* + \Delta G_m)/kT\} \tag{3.8}$$

With increasing supercooling, ΔG^* becomes smaller and thus the rate of nucleation increases. When the system is cooled further, $\Delta G_m \gg \Delta G^*$ and thus the rate of nucleation decreases. The dependence of nucleation rate on temperature is shown in Fig. 3.18.

The nucleation phenomenon discussed above is called homogeneous. When there exist regions such as point defects, dislocations, grain boundaries and surfaces, which tend to decrease interfacial free energy between new and host phases, the activation energy for nucleation decreases and thus nucleation is promoted. This type of nucleation is called heterogeneous. Nucleation phenomena in solids are mostly heterogeneous.

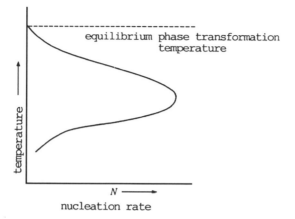

Fig. 3.18 Dependence of nucleation rate on temperature

3.2.4 Crystal Growth

Once nuclei are formed, the nuclei continue to grow as indicated in Fig. 3.17. The system shifts toward equilibrium as the growth of nuclei decreases the total free energy of a system. The growth of nuclei is accommodated by the diffusion of atoms.

Let us assume that the radii of new crystals increase linearly with time once nuclei are formed after an incubation time of t at a certain temperature. As the new crystals grow, phase boundaries also increase at a given speed and eventually touch each other. At this point the growth rate starts to decrease.

The volume of a new phase which grows linearly with time is given by

$$V = (4/3)\pi v^3(t - \tau)^3 \qquad (3.9)$$

where v is the growth rate of phase boundaries. Here it is assumed that the growth of phase boundaries is isotropic. Let $x(t)$ be the volume fraction of a system which has undergone transformation at time t. Then the volume fraction of the system which has not undergone transformation is given by $1 - x(t)$. The volumetric growth rate of a new crystal for a time dt is given by

$$dV_2/dt = \{1 - x(t)\}dV_1/dt$$

The the overall volumetric growth rate is given by

$$dV_3/dt = (Ndt)dV_2/dt = 4\pi v^3 N\{1 - x(t)\}(t - \tau)^2 d\tau \qquad (3.10)$$

where N is the rate of nucleation which is assumed constant.

The volumetric growth rate of new crystals which nucleate between $\tau = 0$ and $\tau = t$ is given by

$$dV_4/dt = dx(t)/dt = 4\pi v^3 N\{1 - x(t)\}\int(t - \tau)^2 d\tau \qquad (3.11)$$

The solution of Eq. (3.11) with an initial condition of $x(0) = 0$ gives

$$x(t) = 1 - \exp\{-(\pi/3)v^3 N t^4\} \qquad (3.12)$$

Avrami performed the analysis in detail and derived the following equation when the nucleation rate is constant:

$$x(t) = 1 - \exp\{-at^n\} \qquad (3.13)$$

where n is called the Avrami index. The constant a contains both thermodynamic and kinetic factors as discussed above. Even if it is possible to demonstrate that experimental data obey Eq. (3.13), the data must be interpreted with caution by using additional information.

3.2.5 Single-crystal growth

The methods to grow single crystals can be classified as indicated in Table 3.2. In order to grow single crystals of a given substance it is necessary to choose an optimal process based on physical properties and a phase diagram of the substance. The gaseous methods in Table 3.2 are used mainly to grow thin films and will be discussed in detail in Section 3.6.2. In this section, the liquid growth methods will be discussed.

It is possible to grow single crystals either by cooling from molten liquids (the melt method) or by dissolving the desired substances in suitable solvents (the solvent method).

Table 3.2 Classification of single-crystal growth methods

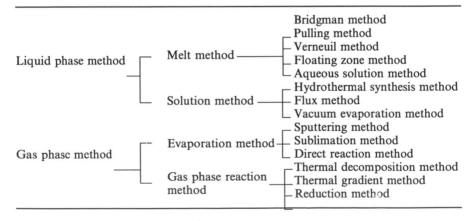

3.2.5.1 THE BRIDGMAN METHOD

In order to grow large single crystals, it is necessary to slow the cooling rate and also to solidify from a single direction. The Bridgman method employs special containers, shown in Fig. 3.19. The container is packed with a powder and heated to a temperature above the melting point of the substance. After the powder is completely molten, the container is gradually cooled to grow a single crystal. On solidification, minute crystals form at the tip of the container. As the container is cooled further, a crystal with a preferential orientation dominates single-crystal growth and a large single crystal appears. One of the fine crystals formed at the tip of the container acts as a seed crystal and thus it is possible to grow a large single crystal. By this method it is possible to grow single crystals of substances which have high vapor pressure at elevated temperatures by sealing the containers. Single crystals of alkali halides, ferrites and $NaNO_2$ are grown by this method.

3.2.5.2 THE PULLING METHOD

This is called either the Czochralski method or the Kyropoulos method. A single crystal is grown by dipping a seed crystal into a melt which is kept at a constant temperature and by pulling it slowly to allow the melt to cool and solidify. In order to avoid dissolution of the seed crystal, it is necessary to cool the seed crystal properly. In addition, one must control the temperature of the melt precisely and pull the growing single crystal correctly. Thus it is possible to grow single crystals with known orientations and very small strains. Single crystals of Si, Ge and alkali halides are grown by this method.

3.2.5.3 THE VERNEUIL METHOD

Using the apparatus shown in Fig. 3.20, powder samples are fed into an oxyhydrogen flame to melt and then to deposit on a seed crystal placed under

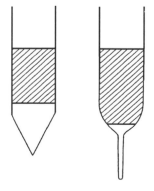

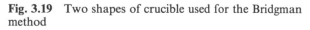

Fig. 3.19 Two shapes of crucible used for the Bridgman method

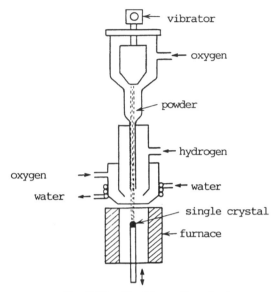

Fig. 3.20 The Verneuil method

the flame. This method does not employ any containers and thus the introduction of impurities can be minimized. Single crystals of ruby, sapphire, spinel, CoO and ferrites are grown by this method.

3.2.5.4 THE FLOATING-ZONE METHOD

As indicated in Fig. 3.21, a polycrystalline rod is melted locally as it rotates. A single crystal results as the molten zone shifts while the rod is slowly moved.

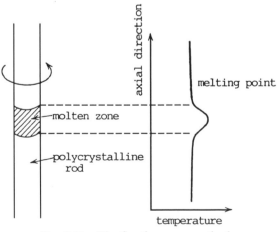

Fig. 3.21 The floating-zone method

Rods are heated by RF, arc imaging or electron beams. Single crystals of ferrites, spinel, ruby, Y_2O_3 and yttrium iron garnet (YIG) are grown by this method.

3.2.5.5 THE AQUEOUS SOLUTION METHOD

Single crystals are grown from supersaturated aqueous solutions. In principle, seed crystals are allowed to grow in conditions where crystal growth occurs without nucleation as indicated by the shaded region in Fig. 3.22. There are three ways to grow single crystals by maintaining a constant supersaturation. In one method water is forced to evaporate as single crystals grow from a solution (the evaporation method). In another method the temperature of an aqueous solution is lowered as single crystals grow from the solution (the slow cooling method). In yet another method raw crystals are dissolved in a high-temperature cell and the resulting aqueous solution is transported to a lower-temperature cell where single crystals grow from the solution. This solution method is called either recirculation or thermal differential. The aqueous solution method is best suited to grow single crystals which have a high solubility in water. Large single crystals of Rochelle salt, potassium diphosphate (KDP) and alum are grown by this method.

3.2.5.6 HYDROTHERMAL SYNTHESIS

This method is a derivative of the aqueous solution method and takes advantage of the fact that although some substances do not dissolve in an

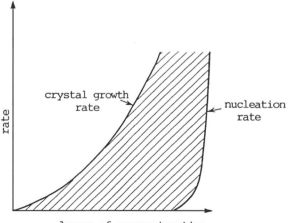

Fig. 3.22 Relationship between the degree of supersaturation and the rate of crystal growth

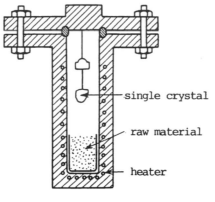

autoclave

Fig. 3.23 The hydrothermal synthesis method

aqueous solution at low temperatures and pressures, they dissolve well in an aqueous solution when heated to elevated temperatures under high pressure. To grow single crystals by this method, it is necessary to have a high-temperature and high-pressure vessel called an autoclave. As indicated in Fig. 3.23, raw crystals are dissolved at the bottom of the autoclave and a single crystal grows onto a seed crystal which is suspended from the top. It is possible to grow single crystals of quartz, silicates, phosphates, ZnO, PbO and ZnS by this method.

3.2.5.7   THE FLUX METHOD

In this method raw crystals are dissolved in a suitable flux and single crystals are grown from a supersaturated solution. This method is suited to substances which vaporize or dissociate at temperatures above their melting points. It is also useful to employ this method to grow single crystals of the substances which do not have suitable containers at elevated temperatures. Single crystals of $BaTiO_3(KF)$, $ZnO(PbF_2)$, $Co_3O_4(B_2O_3-PbO)$, $MgFe_2O_4(NaF)$ are grown by this method. The compounds in the parentheses indicate the fluxes used.

3.2.6   Phase Transformations without Nucleation

In the phase separation of glasses and order–disorder transformations, transformations can occur without nucleation. In the phase separation of glasses this type of phase transformation is called spinodal decomposition. As indicated in Fig. 3.24, the fluctuation of a composition propagates gradually

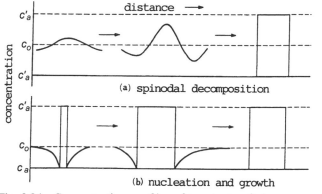

Fig. 3.24 Concentration profiles of two types of phase separation

with time. In the figure a compositional variation of the transformation with nucleation is also shown for comparison.

Phase separation has been observed in a number of glasses, which can be understood as an immiscibility of two liquid phases at intermediate temperatures between liquidus and glass transition temperatures. When the free energy of a solution composed of two components A and B exhibits a temperature dependence shown in Fig. 3.25, the solutions in the shaded regions phase-separate into two solutions with different compositions. This phase separation is called the immiscibility phenomenon. Shaded regions (I) are metastable and region (II) is unstable. Solutions in the metastability regions exhibit immiscibility with droplet-shaped microstructures. On the other hand,

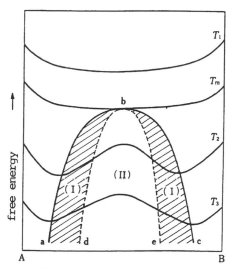

Fig. 3.25 Free energy of a two-component glass and regions of phase separation

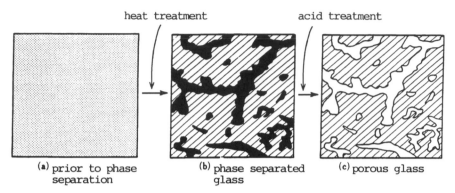

Fig. 3.26 Development of microstructures during the manufacture of a porous glass. (a) Prior to phase separation; (b) phase-separated glass; (c) porous glass

solutions in the unstable region exhibit immiscibility with intertwined microstructures. These observations indicate that in region (I) phase separation is due to nucleation and growth and in region (II) it is the result of spinodal decomposition.

In porous glasses one of the glass phases formed by phase separation is dissolved in an acidic solution. In order to take technical advantage of the phase separation phenomenon, phase-separated glasses must have an intertwined structure of two phases. Furthermore, one of the phases must be easily soluble in aqueous solution and another phase must have a skeletal structure with reasonable mechanical strength. A good example of phase separable glasses is a system of Na_2O–B_2O_3–SiO_2. The phase separable composition of this glass system is quite narrow and is given by $Na_2O:B_2O_3:SiO_2 = 5$–$8:20$–$25:70$–75 wt%. When the melts are heat treated between 500°C and 600°C, the melts separate into SiO_2-rich and Na_2O–B_2O_3-rich phases. Porous glasses result after the Na_2O–B_2O_3-rich phase is dissolved in an acidic solution such as H_2SO_4. A compositional change due to phase separation is shown schematically in Fig. 3.26.

3.3 DIFFUSION AND MASS TRANSPORT IN SOLIDS

When a cube of sugar is dropped into a cup of coffee the sugar dissolves into the coffee and after a while its concentration becomes uniform. This homogenizing phenomenon by mass transport in a solvent is called diffusion.

In general, atoms, molecules and ions are constantly in irregular thermal motion. When viewed macroscopically, a substance in equilibrium looks motionless, but when viewed microscopically, particles which constitute the substance are constantly in motion. Diffusion due to thermal motion alone is called self-diffusion. The diffusion of a solute from a region of high concentration to one of low concentration under a concentration gradient as in the case of diffusion of sugar molecules in coffee is called either chemical diffusion or interdiffusion.

3.3.1   The Laws of Diffusion

As a simple example, let us consider the heating of two pure metals A and B which are in intimate contact. Prior to the onset of diffusion, no A atoms can be found in B and vice versa. After the onset of diffusion, A atoms diffuse into B and B atoms diffuse into A. Finally the pair forms a homogeneous alloy. This diffusion phenomenon is shown schematically in Fig. 3.27. Fick derived two equations to treat diffusion phenomena mathematically.

Let us consider diffusion in one dimension. A concentration gradient dc/dt along the x direction creates a flux J per unit area per unit time. Since the flux is proportional to the concentration gradient, J can be given by

$$J = -D(dc/dt) \tag{3.14}$$

where D is the diffusion coefficient. Equation (3.14) is called Fick's first law. The negative sign in Eq. (3.14) is due to the fact that a substance diffuses opposite to the concentration gradient.

In actual diffusion it is common to see the time variation of a concentration. In these cases the concentration is given by a function of both position x and time t. The time variation of a concentration at a certain position can be given by

$$\partial c/\partial t = \partial/\partial x\{D(\partial c/\partial x)\} \tag{3.15}$$

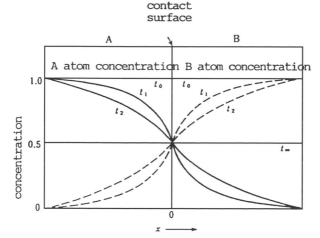

Fig. 3.27   Time variation of atomic concentration profiles of a diffusion pair of metals A and B

This equation is called either Fick's second law or the diffusion equation. When the diffusion coefficient does not depend on concentration, Eq. (3.15) can be simplified as follows:

$$\partial c/\partial t = D(\partial^2 c/\partial x^2) \tag{3.16}$$

By solving this equation with initial and boundary conditions, it is possible to understand the progress of a diffusion process.

Let us consider a concentration profile when a pair of A and B metals are in contact at elevated temperatures. Let c be the concentration of metal A. Let initial conditions be as follows:

$$c = c_0 \text{ for } x < 0$$
$$c = 0 \text{ for } x > 0$$

The solution of Eq. (3.16) is given by

$$c(x, t) = \frac{c_0}{2}\left\{1 - \frac{2}{\sqrt{\pi}}\int_0^{\frac{x}{2\sqrt{Dt}}} e^{-y^2} dy\right\} \tag{3.17}$$

The second term in the right-hand side of the equation is called the error function and is defined as follows:

$$\text{erf}(z) = [2/(\pi)^{1/2}]\int_0^z \exp(-y^2) dy \tag{3.18}$$

The error function has the following characteristics:

$$\text{erf}(-z) = -\text{erf}(z)$$
$$\text{erf}(\infty) = 1 \tag{3.19}$$
$$\text{erf}(0) = 0$$

Equation (3.17) can be rewritten with an error function as follows:

$$c(x, t) = (c_0/2)\{1 - \text{erf}(x/2(Dt)^{1/2}\} \tag{3.20}$$

The concentration of metal A is shown by the solid curves in Fig. 3.27.

Let us consider the diffusion of a radioisotope which is deposited on a surface. Self-diffusion coefficients can be obtained experimentally by conducting this type of diffusion experiment. The solution of Eq. (3.16) with an initial condition of $c = 0$ for $x > 0$ at $t = 0$ gives

$$c = (c_0/[\pi Dt]^{1/2})\exp(-x^2/4Dt) \tag{3.21}$$

As shown in Fig. 3.28, the concentration profiles represented by Eq. (3.21) have a Gaussian distribution. As indicated in Fig. 3.29, $\ln(c/c_0)$ has a linear relationship with x^2 and thus it is possible to determine D from its slope. In general, diffusion coefficients depend on concentration. Hence in order to

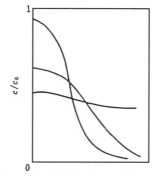

Fig. 3.28 Gaussian distributions of a solute concentration

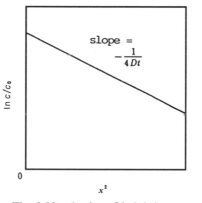

Fig. 3.29 A plot of $\ln(c/c_0)$ versus x^2

determine concentration profiles as a function of time and position, it is necessary to solve Eq. (3.15) instead of Eq. (3.16).

3.3.2 Random Walk and Self-diffusion

Like diffusion in gases and liquids by irregular thermal motion of molecules (Brownian motion), diffusion in solids can be assumed to take place by random walks. Let us consider the random walks of a radioisotope (a tracer) in a simple cubic lattice as shown in Fig. 3.30. Let f be the jump frequency of the tracer from one site to another per unit of time. When the concentrations of the tracer on two crystal planes, 1 and 2, which are a distance r apart, are n_1 and n_2, the number of tracer atoms which jump from 1 to 2 for δt is given by $(1/6)n_1 f \delta t$ and the number of tracer atoms which jump from 2 to 1 for δt is $(1/6)n_2 f \delta t$. Here it is assumed that tracer atoms have an equal jump probability in the x, y and z directions. The flux J of tracer atoms from 1 to 2 is given by

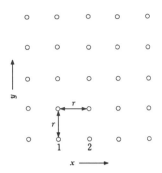

Fig. 3.30   Schematic of an atomic arrangement

$$J = (1/6)(n_1 - n_2)f \qquad (3.22)$$

The tracer concentrations per unit volume are given by

$$c_1 = n_1/r \quad c_2 = n_2/r \qquad (3.23)$$

Equations (3.22) and (3.23) with $c_2 - c_1 = r(dc/dt)$ give

$$J = -(1/6)fr^2(dc/dt) \qquad (3.24)$$

This equation is equivalent to Eq. (3.14) (Fick's first law) and thus

$$D = (1/6)fr^2 \qquad (3.25)$$

The values of f and r depend on a specific diffusion mechanism which will be discussed in the following section. Thus it is possible to understand the diffusion of atoms in solids by analyzing both f and r in detail.

3.3.3   Diffusion in Crystals

The diffusion of atoms in crystals which contain defects occurs by the motion of the defects and diffusion takes place by vacancy, interstitialcy or semi-interstitialcy mechanisms. In addition, diffusion can, in principle, take place by direct-exchange and ring mechanisms the existence of which has not been confirmed experimentally. These diffusion mechanisms are shown schematically in Fig. 3.31. As discussed above, the relationship of a diffusion coefficient with a jump frequency and distance is given by Eq. (3.25).

The change in free energy of an atom which jumps from one lattice site to an adjacent vacant site is shown schematically in Fig. 3.32. Atoms with ΔG_m can overcome the barrier for diffusion and can jump to the adjacent vacant sites. The jump frequency depends on the vibration frequency of an atom v, the Boltzmann factor and the probability of having a vacancy at the adjacent sites, which is the vacancy concentration N_v and is given by

$$f = AvN_v \exp(-\Delta G_m/RT) \qquad (3.26)$$

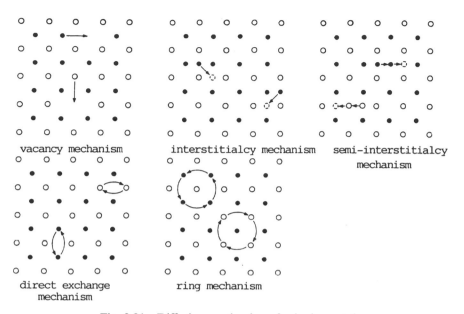

Fig. 3.31   Diffusion mechanisms for ionic crystals

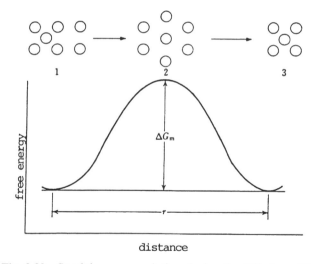

Fig. 3.32   Spatial energy variation during the diffusion of ions

where A is the constant which depends on the crystal structure. By substituting Eq. (3.26) into Eq. (3.25), a self-diffusion coefficient by the vacancy mechanism can be given by

$$D = (1/6)Avr^2N_v \exp(\Delta S_m/T)\exp(-\Delta H_m/RT) \qquad (3.27)$$

The vacancy concentration N_v is the sum of vacancies at thermal equilibrium n_v and vacancies introduced by impurities n_i. Thus Eq. (3.27) can be rewritten to

$$D = (1/6)Avr^2(n_v + n_i)\exp(\Delta S_m/T)\exp(-\Delta H_m/RT) \qquad (3.28)$$

At low temperatures vacancies introduced by impurities predominate, but vacancies at thermal equilibrium predominate at high temperatures. The temperature range at which vacancies introduced by impurities predominate is called the extrinsic region and the self-diffusion coefficient at this range can be approximated by

$$D = (1/6)Avr^2 n_i \exp(\Delta S_m/T)\exp(-\Delta H_m/RT) \qquad (3.29)$$

Since n_i is a constant which depends on species and quantities of impurities present, the temperature dependence of the self-diffusion coefficient arises mainly from ΔH_m. The high-temperature range at which the vacancy concentration is not governed by impurities is called the intrinsic region. The self-diffusion coefficient of crystals with Schottky defects is given by

$$D = (1/6)Avr^2 \exp([\Delta S_m + \Delta S_f/2]/T)\exp(-[\Delta H_m + \Delta H_f/2]/RT) \qquad (3.30)$$

where ΔH_f and ΔS_f are the enthalpy and entropy of the defect formation, respectively. In this case the temperature dependence of D comes from $\Delta H_m + \Delta H_f/2$.

For the interdiffusion of A and B atoms, the chemical diffusion coefficient, D, can be given by

$$D = (D_A N_B + D_B N_A)(1 + d\ln\gamma_A/d\ln N_A) \qquad (3.31)$$

where γ_A is the activity coefficient of A atoms, D_A and D_B are the self-diffusion coefficients of A and B and N_A and N_B are the concentrations of A and B, respectively.

In general, the diffusion coefficient can be expressed by the Arrhenius relationship:

$$D = D_0 \exp(-Q/RT) \qquad (3.32)$$

It is necessary to exercise utmost caution in interpreting the significance of the activation energies which are determined experimentally.

The self-diffusion coefficient of Na^+ in NaCl is shown in Fig. 3.33. It is known that NaCl contains the Schottky defects. It is also evident from the fact that ionic radii of Na^+ and Cl^- are 1.02 Å and 1.81 Å, respectively, that Na^+ ions are significantly more mobile than Cl^- ions. The high- and low-temperature regions of the self-diffusion coefficient observed correspond to the intrinsic and extrinsic regions, respectively. From the slopes, activation energies are calculated as 41.5 kcal/mol ($= \Delta H_m + \Delta H_f/2$) and 17.7 kcal/mol ($= \Delta H_m$), respectively. Thus H_f is 47.6 kcal/mol. Formation enthalpies of the

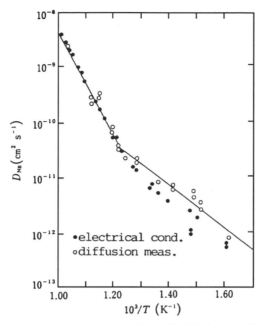

Fig. 3.33 Temperature dependence of the self diffusion coefficient of Na^+ ions in NaCl. (Reproduced by permission of The American Institute of Physics, from D. Mapother, H. N. Crooks and R. Maurer, *J. Chem. Phys.*, **18**, 1231, 1950)

Schottky defects and activation enthalpy for cation motion of various halides are determined similarly and are listed in Table 3.3.

Defect concentrations of non-stoichiometric compounds depend on oxygen partial pressure. For example, it is known that CoO, which is a cation-deficient oxide, contains Co vacancies and its defect formation reactions can be expressed as follows:

Table 3.3 Formation enthalpy of Schottky defects, ΔH_f and activation energy for diffusion, ΔH_m of alkali halides

Halide	ΔH_f (kcal mol^{-1})	ΔH_m (kcal mol^{-1})	Halide	ΔH_f (kcal mol^{-1})	ΔH_m (kcal mol^{-1})
LiF	53.9	16.1	KCl	59.9	16.4
LiCl	48.9	9.22	CsCl	42.9	13.8
LiBr	41.5	8.99	CsBr	46.1	13.4
NaCl	47.6	17.7	CsI	43.8	13.4
NaBr	38.7	18.4	TlCl	30.0	11.5

W. D. Kingery, H. K. Bowen and D. R. Uhlmann, *Introduction to Ceramics*, 2nd edition, p. 236, John Wiley (1976). ©1976 John Wiley.

$$(1/2)O_2(g) = V_{Co}^x + O_O^x \quad \Delta H_1, K_1$$
$$V_{Co}^x = V_{Co}' + h^{\cdot} \quad \Delta H_2, K_2 \quad (3.33)$$
$$V_{Co}' = V_{Co}'' + h^{\cdot} \quad \Delta H_3, K_3$$

It is known that the majority of defects at temperatures higher than 900°C are dictated by oxygen partial pressure. V_{Co}' are predominant at $P_{O_2} >\sim 10^{-4}$ atm and V_{Co}'' at $P_{O_2} <\sim 10^{-5}$ atm. Application of the mass action rule to reactions (3.33) gives the following relationships:

$$N_v \approx [V_{Co}'] = (K_1 K_2)^{1/2} P_{O_2}^{1/4} \text{ for high } P_{O_2}$$
$$N_v \approx [V_{Co}''] = \{(1/4)K_1 K_2 K_3\}^{1/3} P_{O_2}^{1/6} \text{ for low } P_{O_2} \quad (3.34)$$

The substitution of Eqs (3.34) into Eq. (3.27) yields the relationships of D with P_{O2} and T, which are shown schematically in Fig. 3.34.

3.3.4 Grain Boundary and Surface Diffusion

The diffusion which has been discussed thus far takes place by the transport of atoms via point defects such as vacancies and interstitials and is called either grain matrix diffusion or internal diffusion. It is known that diffusion also occurs via structural defects such as dislocations and grain boundaries. In general, the rates of diffusion depend on the types of diffusion involved. Diffusion of atoms via grain boundaries is called grain boundary diffusion and that on surfaces is surface diffusion.

Surfaces have more distorted atomic arrangements than grain boundaries which in turn have more distorted atomic arrangements than grain matrices.

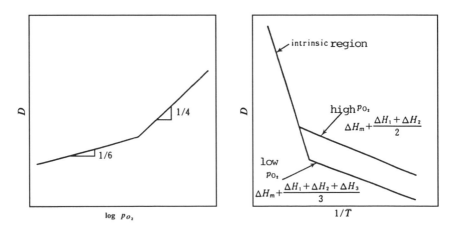

Fig. 3.34 Dependence of diffusion coefficients on oxygen partial pressure and temperature

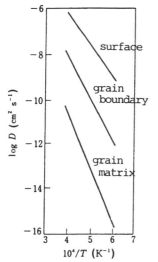

Fig. 3.35 Diffusion coefficients of thorium in tungsten. (From T. Obara, *Treatise of Microstructures in Metals*, Asakura Publishing Co., 1971, p. 57)

Thus it is expected that the transport of atoms and ions by diffusion increases in the following order:

$$\text{Grain matrix} < \text{grain boundary} < \text{surface}$$

The diffusion coefficients of thorium in tungsten are shown in Fig. 3.35. In the figure it is also evident that the higher the diffusion coefficients, the lower the activation energies.

The self-diffusion coefficients of Al^{3+} and O^{2-}, D_{Al} and D_O in α-Al_2O_3 are shown in Fig. 3.36. Although D_{Al} are identical in both single-crystal and polycrystalline samples, D_O of a polycrystalline sample is higher than that of a single-crystal sample, which is attributed to the enhancement of O^{2-} diffusion via grain boundaries. The enhancement of diffusion via grain boundaries has been observed for O^{2-} in MgO and for U^{4+} in UO_2. On the other hand, the enhancement of diffusion via grain boundaries has not been observed for Mg^{2+} in MgO and for O^{2-} in UO_2. Thus the enhancement of diffusion in the order of grain matrix < grain boundary < surface, which is almost universally applicable to metals, is not necessarily valid for all ions in ionic crystals.

3.3.5 Solid-state Reactions

A system which is not in thermodynamic equilibrium has a tendency to shift to an equilibrium condition whose free energy is at minimum. It is possible from a phase diagram to identify the equilibrium condition of a system at a certain temperature and pressure.

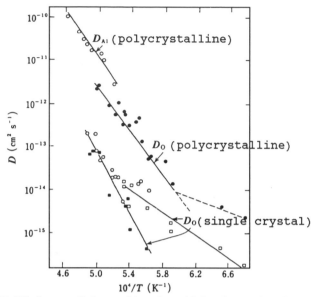

Fig. 3.36 Self-diffusion coefficients of ions in α-Al$_2$O$_3$. (Reproduced by permission of The American Institute of Physics, from Y. Oishi and W. D. Kingery, *J. Chem. Phys.*, **33**, 905, 1960)

Various reactions take place as the system tries to shift toward equilibrium. It is possible to determine from thermodynamic functions of reactants and products whether a given reaction will take place or not. Thus thermodynamically, the driving force of a reaction is the difference in chemical potential between reactants and products. On the other hand, the kinetic and mechanistic information of a reaction has to be obtained from detailed analysis of experimental data.

Solid-state reactions imply reactions in solids which do not involve any gas or liquid during reaction. The transport of atoms and ions in solids is very slow and thus solid-state reactions are also very slow. Reaction rates depend on the rate-controlling steps of the reactions, but regardless of the rate-controlling steps the rate constant for a reaction which involves atoms and ions can be expressed by the following Arrhenius equation:

$$k = k_0 \exp(-Q/RT) \tag{3.35}$$

where Q is the activation energy of a reaction. Since Q is always positive, the reaction rate increases with increasing temperature.

3.3.5.1 COMBINATION REACTIONS

When a pair of MgO and Al$_2$O$_3$, either single crystal or polycrystalline, is heated to 1400°C, MgAl$_2$O$_4$ spinel forms between them as indicated in

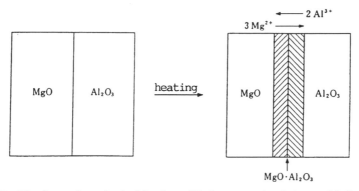

Fig. 3.37 The formation of spinel by the solid phase reaction between MgO and Al_2O_3

Fig. 3.37. As expected from the phase diagram shown in Fig. 3.38, the thickness of the spinel layer x increases with time t and is given by:

$$x^2 = k_1 t \qquad (3.36)$$

where k_1 is the rate constant which can be expressed by Eq. (3.35). Equation (3.36) is called the parabolic law and is valid for the reactions which are rate-controlled by the diffusion of atoms and ions. While maintaining electrical neutrality locally, the spinel phase forms by the diffusion of Al^{3+} ions toward MgO and Mg^{2+} ions toward Al_2O_3. Similar combination reactions have been observed in systems which have intermediate compounds such as $MgO–TiO_2$ and $CaO–SiO_2$.

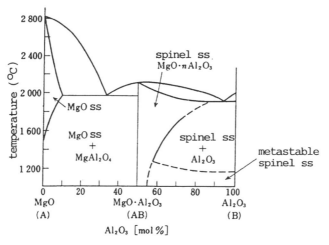

Fig. 3.38 Phase diagram of the $MgO–Al_2O_3$ system. (From Japan Chemical Society (ed.), *Treatise of Chemistry IX: Inorganic Reactions with Solids*, Society Publications Center, 1975, p. 87: reproduced with permission)

3.3.5.2   SOLID SOLUTION AND DISSOLUTION–PRECIPITATION REACTIONS

The MgO–NiO system exhibits complete solid solubility. A homogeneous solid solution results by interdiffusion when a pair of MgO and NiO in contact is heated at high temperatures. This type of solid-state reaction is called solid solution.

When a pair of $MgO \cdot Al_2O_3$ and Al_2O_3 in contact is heated to elevated temperatures, the spinel dissolves Al_2O_3 to its solubility limit and thus forms a solid solution whose chemical formula can be expressed by $MgO \cdot nAl_2O_3$. Here n depends on temperature. When the solid solutions $MgO \cdot nAl_2O_3$ are cooled to lower temperatures where two phases co-exist, the excess Al_2O_3 precipitates out from the solid solution. This type of solid-state reaction is called dissolution–precipitation.

3.3.5.3   DEVITRIFICATION REACTIONS

When a transparent glass of SiO_2 formed by quenching is heated to around 1100°C for an extended period of time, the glass becomes opaque due to the precipitation of β-cristobalite. The reaction which involves opacification of transparent glasses is called devitrification. When heated, soda lime glasses which contain F also devitrify due to the precipitation of NaF or CaF_2.

Glass ceramics which contain both crystalline and glassy phases are produced by devitrification reactions. When a few weight per cent of nucleation promoters with fine grain sizes ($< 1 \, \mu m$) such as TiO_2 and ZrO_2 are added to glasses, high-strength materials with very little porosity can be obtained. Today it is possible to manufacture clear glass ceramics, and thus it has become incorrect to call this type of reaction devitrification.

3.4   SOLID–GAS, SOLID–LIQUID REACTIONS

In this section reactions between solid and gaseous phases and between solid and liquid phases will be discussed. Even at room temperature solids react with air and moisture (such as oxidation) as well as vaporize, although the rates are extremely small. At elevated temperatures these reactions proceed much faster. In addition, it is also possible to see the decomposition of solids, which generates some gaseous species (thermal decomposition). Gaseous reactions at elevated temperatures which result in the formation of solids are very important for the syntheses of fine powders and thin films. Furthermore, solid–liquid reactions include hydration reactions of cements, melt–solidification reactions and precipitation reactions from aqueous solutions.

3.4.1 Evaporation Reactions

There are two types of solid evaporation: homogeneous and non-homogeneous. When a solid evaporates without a change in solid composition, the reaction is called homogeneous evaporation. Homogeneous reactions can be expressed as follows:

$$M(s) \rightarrow M(g), M_2(g), M_3(g),. . .$$

$$MX(s) \rightarrow MX(g), M(g), X_2(g),. . .$$

Solids vaporize partially or wholly. Gaseous species generated differ from one substance to another and can be identified by mass spectrometers. Examples of homogeneous gaseous reactions are listed in Table 3.4.

Some components tend to vaporize preferentially from compounds which have multiple components. This type of evaporation is called non-homogeneous. As vaporization proceeds, the surface composition of a compound changes with time. Examples of preferential vaporization are Na_2O from soda lime glasses, PbO from lead glasses and PZT[Pb(Zr, Ti)O_3], ZnO from $ZnAl_2O_4$ and CaO from calcia-stabilized zirconia.

Consider the preferential vaporization of a component from a homogeneous substance, illustrated schematically in Fig. 3.39. At onset of vaporization decomposition occurs at the surface. The vaporizable component becomes a gas and is transported away from the surface by gaseous diffusion. After initial vaporization, the gaseous component has to diffuse through a layer of the solid reaction product (solid phase II in Fig. 3.39(b)) and vaporizes at the surface. The vaporization is rate-controlled by decomposition, by diffusion of the gaseous species through the solid reaction product, or by vaporization at the surface. In general, the gaseous diffusion of the vaporizable species is swift and thus does not influence vaporization processes.

When vaporization is rate-controlled by the decomposition reaction, the vaporization rate is constant with time and thus the amount vaporized per unit surface area increases linearly with time. For example, Cr_2O_3 vaporizes

Table 3.4   Examples of evaporation reactions

Compound	Vapor species
Li_2O	$\rightarrow Li + Li_2O(1 \sim 10\%) + LiO(0.1\%) + Li_2O$
MgO	$\rightarrow Mg + O + MgO$ (very little)
PbO	$\rightarrow Pb + PbO + (PbO)_2 + (PbO)_4$
Al_2O_3	$\rightarrow Al + O + AlO + Al_2O + O_2$
SiO_2	$\rightarrow SiO + O_2 + SiO_2$
SiC	$\rightarrow Si + SiC_2 + SiC$
Si_3N_4	$\rightarrow Si + N_2$

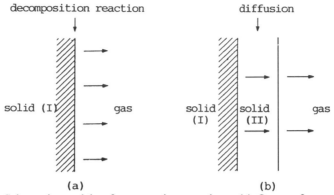

Fig. 3.39 Schematic models of evaporation reactions. (a) Onset of evaporation; (b) after the formation of a solid product layer

preferentially from $LaCrO_3$. Because of the formation of a porous La_2O_3 reaction product, vaporization is rate-controlled by the decomposition reaction of $LaCrO_3$.

When vaporization is rate-controlled by the diffusion in a solid, the vaporization rate decreases with time and the amount vaporized is proportional to one half power of time, which is the parabolic law. The vaporization of MgO from $MgO \cdot nAl_2O_3$ is rate-controlled initially by the decomposition reaction, but once the solid surface is covered with a dense layer of Al_2O_3, the vaporization is rate-controlled by diffusion. The time dependence of the MgO vaporization from $MgO \cdot nAl_2O_3$ is shown in Fig. 3.40.

3.4.2 Oxidation Reactions

As an example of oxidation reactions of metals, let us consider the oxidation of Ni. The resulting NiO surface layer is commonly called scale. While a layer of NiO is forming, the equilibrium oxygen partial pressure at the interface between NiO and O_2 is that of the surrounding atmosphere, $p_{O_2}(I)$. The equilibrium oxygen partial pressure at the interface between Ni and NiO is that of the decomposition pressure of NiO, $p_{O_2}(II)$. Thus as indicated in Fig. 3.41, there exists a gradient of oxygen partial pressure (chemical potential of oxygen) in the layer of NiO.

NiO is a metal-deficient non-stoichiometric compound. Since the concentrations of cation vacancies and electron holes depend on p_{O2}, the concentrations of V_{Ni}'' and $h^{\cdot}$ exhibit gradients, as indicated in the figure, due to the gradient of oxygen partial pressure in NiO. As a consequence, both V_{Ni}'' and $h^{\cdot}$ diffuse from the surface to the interface between Ni and NiO. On the other hand, Ni^{2+} and electrons migrate toward the surface. Thus the growth of the NiO scale obeys the parabolic law.

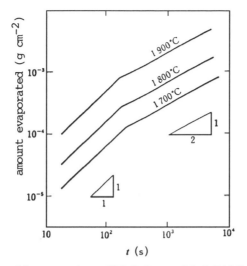

Fig. 3.40 Preferential evaporation of MgO from a MgO·8Al$_2$O$_3$ single crystal. (From T. Sata and T. Yokoyama, *Yogyo Kyokai Shi*, **81**, 170 (1973)

Oxidation reactions are important not only in metals but also in non-oxide ceramics. Let us consider the oxidation of silicon nitride, Si$_3$N$_4$, at high temperatures as an example. It is known that the oxidation of silicon nitride exhibits the following two types of reaction:

(I) : $2Si_3N_4 + 3O_2 = 6SiO(g) + 4N_2(g)$ at low oxygen partial pressures

(II) : $Si_3N_4 + 3O_2 = 3SiO_2(s) + 2N_2(g)$ at high oxygen partial pressures

Reaction (I) is called active oxidation and reaction (II) passive oxidation. Since reaction (I) generates gaseous products only, the oxidation proceeds violently. On the other hand, reaction (II) forms an oxide layer of SiO$_2$ on the surface

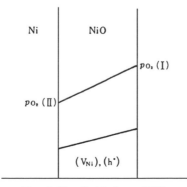

Fig. 3.41 Oxidation of Ni

and thus the oxidation rate decreases with time. When the oxidation is rate-controlled by the diffusion of oxygen in the oxide film it is possible to see the growth of an oxide scale which obeys the parabolic law. However, the oxidation of Si_3N_4 is known to be influenced by the microstructure and by the presence of sintering additives, and thus is very complicated.

3.4.3 Thermal Decomposition Reactions

When heated to high temperatures, carbonates, oxalates, nitrates and hydroxides decompose and generate gases. This type of reaction is called thermal decomposition.

A combination of differential thermal analysis (DTA) and thermogravimetry (TG) has been used extensively for the study of thermal decomposition. Let us consider the dehydration reaction of one of the aluminum hydroxides, gibbsite ($Al_2O_3 \cdot 3H_2O$). DTA–TG curves of two gibbsite raw materials, one with coarse grains (50–60 μm) and the other with fine grains ($< 1\ \mu$m), heated at a rate of 10°C/min, are shown in Fig. 3.42. The coarse grains of gibbsite lose absorbed water at around 200°C, undergo a main dehydration reaction at around 300°C and lose a small amount of additional water at around 550°C. On the other hand, only the main dehydration reaction at around 300°C can be observed in the fine grains of gibbsite. For comparison, the dehydration of boehmite ($Al_2O_3 \cdot H_2O$) at around 550°C is also shown in Fig. 3.42. This dehydration peak coincides with the third peak of the coarse grains of gibbsite. Thus it was anticipated that coarse gibbsite forms a small amount of boehmite during

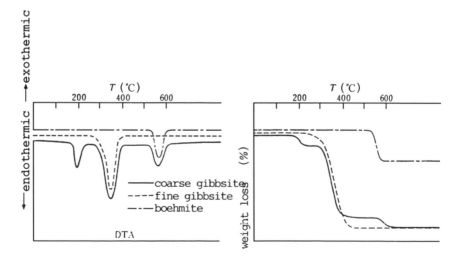

Fig. 3.42 Dehydration reactions of aluminum hydroxides

dehydration. Structural analyses by X-ray diffraction have confirmed the presence of the following dehydration processes:

(1) Coarse gibbsite
$$\nearrow \text{boehmite} \rightarrow \gamma\text{-Al}_2\text{O}_3$$
$$\searrow \chi\text{-Al}_2\text{O}_3$$

(2) Fine gibbsite $\rightarrow \chi$-Al$_2$O$_3$
(3) Boehmite $\rightarrow \gamma$-Al$_2$O$_3$

It is known that the formation of boehmite from gibbsite is catalyzed by water.

In general, thermal decomposition of solids initiates at some structural defects such as surfaces, grain boundaries and dislocations. At the onset of thermal decomposition, small nuclei are formed. Subsequently these nuclei grow with time. Thermal decomposition at a constant temperature proceeds as indicated in Fig. 3.43. Reaction (a) exhibits an incubation period with very little weight change, which is followed by a steady increase in weight loss. This type of thermal decomposition has been observed in $NiC_2O_4 \cdot 2H_2O$. Reaction (b) is the most common type of thermal decomposition and is called the sigmoid. Initially a large number of nuclei are formed in a solid. Reaction fronts expand with the growth of the nuclei. When resulting grains touch each other, the reaction rate starts to decrease. This type of thermal decomposition has been observed in Co_3O_4 and $MgCO_3$ and can be represented by the Avrami–Erofeev equation, which is given by

$$\alpha = 1 - \exp(-at^l) \tag{3.37}$$

where α is the amount of a substance decomposed and a and l are the material-specific constants which also depend on temperature.

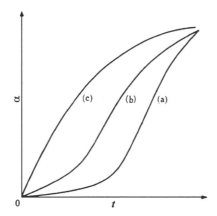

Fig. 3.43 The amount decomposed, α, as a function of time in various thermal decomposition reactions

Reaction (c) represents thermal decomposition with a decreasing reaction rate. This type of thermal decomposition can be observed either in cases where a large number of nuclei are formed on the surface and grow inward with decreasing area of reaction fronts or where individual grains decompose at a constant probability. The former cases can be represented by Mampel's equation,

$$1 - (1 - \alpha)^{1/n} = kt/r_0 \tag{3.38}$$

where k is the rate constant and r_0 the initial radius of grains. The constant n is 3 for spherical grains and 2 for cylindrical grains. This type of thermal decomposition has been observed in $CaCO_3$ and $Mg(OH)_2$. The latter case is called a primary reaction and can be represented by

$$d\alpha/dt = k(1 - \alpha) \tag{3.39}$$

This type of thermal decomposition has been observed in the dehydration and decomposition of fine particles.

3.4.4 Gas Phase Reactions

Chemical reactions which involve the formation of solids from gases are called gas phase reactions. The formation of solids from gases alone is called chemical vapor deposition (CVD). The transport of solids by gaseous carriers is called chemical vapor transport (CVT). The formation of fine powders and thin films by CVD is one of the most important manufacturing processes in ceramic sciences.

3.4.4.1 CHEMICAL VAPOR DEPOSITION (CVD)

Two general types of CVD processes are known. One is the formation of a solid by the thermal decomposition of a homogeneous gas, which can be represented by $A(g) \rightarrow B(s) + C(g)$. The other type is the formation of a solid by the chemical reaction of two or more gaseous species, which can be represented by $A(g) + B(g) \rightarrow C(s) + D(g)$. Examples of the former CVD process are as follows:

$$CH_3SiCl_3(g) \rightarrow SiC(s) + 3HCl(g)$$
$$CH_4(g) \rightarrow C(s) + 2H_2(g)$$

However, in general, the CVD method refers to the latter process. A variety of gaseous reactions can be employed in CVD processes. Some examples are as follows:

- Hydration–decomposition:

$$2AlCl_3(g) + 3H_2O(g) \rightarrow Al_2O_3(s) + 6HCl(g)$$

- Oxidation–reduction:

$$AlCl_3(g) + (1/2)N_2 + (3/1)H_2 \rightarrow AlN(s) + 3HCl(g)$$

- Substitution:

$$TiCl_4(g) + O_2 \rightarrow TiO_2(s) + 2Cl_2(g)$$

In order to promote CVD reactions it is necessary to activate gas molecules by providing additional energies. Extra energies which are required to enhance the thermal motion of molecules, the excitation and the formation of radicals can be provided by heating, laser irradiation, plasmas, etc.

Various morphologies of solids, which include fine powders, whiskers, single crystals and polycrystals, can be produced by CVD (Fig. 3.44). It is well known that, depending on the combination of gaseous species, the reaction temperature, the flow rate and the mode of activation, it is possible to alter the morphology of a solid formed. As indicated in Fig. 3.45, the degree of supersaturation of a reaction system is the main cause of the formation of various morphologies. The degree of supersaturation is defined by the ratio of an actual solid vapor pressure to an equilibrium vapor pressure of a solid precipitated. For example, let us consider the following CVD reaction:

$$TiCl_4(g) + CH_4(g) \rightarrow TiC(g) + 4HCl(g)$$

The degree of supersaturation for this reaction is given by P/P_0 where P and P_0 are the partial pressure of TiC(g) generated by the reaction and the equilibrium vapor pressure of TiC(s) at the reaction temperature, respectively. With increasing P/P_0 the deviation from equilibrium increases, which results in more frequent nucleation than crystal growth. As a consequence, the solid

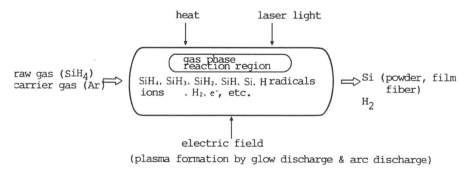

Fig. 3.44 Principle of the CVD method (formation of Si by the decomposition of SiH$_4$). (From S. Mizuta and K. Kawamoto, *Materials Technology Vol. 13, Ceramic Materials*, Tokyo University Press, 1986, p. 168: reproduced with permission)

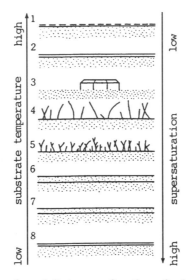

Fig. 3.45 Morphology of precipitates as a function of substrate temperature and the degree of supersaturation. 1 Etching, 2 epitaxial growth, 3 platelet growth, 4 whisker growth, 5 growth of needles, 6 columnar growth, 7 fine crystallites, 8 amorphous solids

morphology of a homogeneous nucleation system changes as follows: single crystal → polycrystalline → coarse powder → fine power → amorphous. The change in solid morphology of a heterogeneous nucleation system on a substrate is shown schematically in Fig. 3.45. In addition to supersaturation, the solid morphology is also affected by the rate of a chemical reaction.

3.4.4.2   CHEMICAL VAPOR TRANSPORT (CVT)

When a new gas species is formed by the reaction between a solid and a gas and is transported to another location where a solid phase precipitates from the gas phase, this is called chemical vapor transport. In an example shown in Fig. 3.46, the following reaction takes place at the high-temperature end of a closed tube:

$$Fe_2O_3(s) + 6HCl(g) = 2FeCL_3(g) + 3H_2O(g) \tag{3.40}$$

Both $FeCl_3$ and H_2O are transported to the low-temperature end of the closed tube where a reverse reaction forms Fe_2O_3 crystals together with HCl. HCl is transported back to the high-temperature end and the cycle repeats continuously. In the figure, CVT with a closed tube is shown, but in some applications it is also possible to conduct CVT in an open tube. The equilibrium constant of reaction (3.40) can be given by

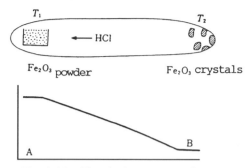

Fig. 3.46 Chemical transport of Fe_2O_3

$$K = p^2_{FeCl_3} p^3_{H_2O} / p^6_{HCl} = \exp(\Delta H / RT) \exp(\Delta S / R) \qquad (3.41)$$

where ΔH and ΔS are the enthalpy and entropy changes of this reaction. Since reaction (3.40) is endothermic ($\Delta H > 0$), the vapor pressures of $FeCl_3$ and H_2O are higher at high temperatures and thus the transport occurs from the high-temperature end to the low-temperature end. However, transport in an exothermic reaction ($\Delta H < 0$) occurs from low to high temperatures. It is not possible to carry out CVT by a chemical reaction with $\Delta H = 0$. Additional examples of CVT are as follows:

(1)   Transport from high to low temperatures:

$$Ni(s) + 2HCl(g) = NiCl_2(g) + H_2(g)$$
$$CdS(s) + I_2(g) = CdI(g) + (1/2)S_2(g)$$

(2)   Transport from low to high temperatures:

$$SiO_2(s) + 4HF(g) = SiF_2(g) + 2H_2O(g)$$
$$Zr(s) + 2I_2(g) = ZrI_4(g)$$

CVT processes have been used commercially for the syntheses of single crystals, fine powders and thin films and the purification of many substances.

3.4.5   Cement Production

The process flow for cement production and a schematic layout of a cement plant are shown in Figs 3.47 and 3.48, respectively. Two main ingredients of a cement, calcite and clay with additives such as soft siliceous stone, sandstone and slug, are fed at a constant speed to a tube mill for mixing and grinding. After grinding, the raw powders are fed to a suspension preheater and preheated to 800°C by the exhaust gases from a rotary kiln. The preheated powders are fed to a slowly rotating kiln and heated to a maximum temperature of 1600°C. Thermal decomposition and solid-state reactions

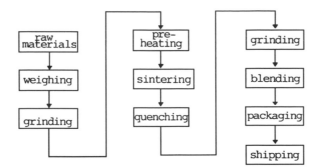

Fig. 3.47 Process flow for the manufacture of cements

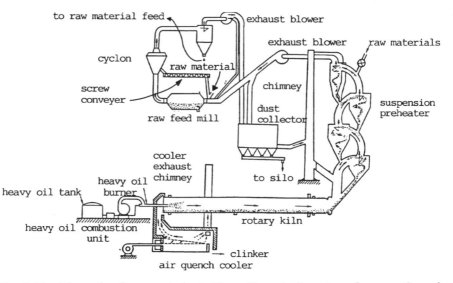

Fig. 3.48 Schematic of a cement plant. (From *Ceramic Operations*, Japanese Ceramic Society, 1971, p. 202: reproduced with permission)

take place during heating and a sintered product called clinker is produced. The clinker is quenched to 50–60°C at the exit of the rotary kiln. In order to control the hydration speed, the clinker is mixed and ground with about 3.5% of plaster ($CaSO_4 \cdot 2H_2O$). The production of cements involves complicated chemical reactions during heating including thermal decomposition and solid-state reactions.

The solidification of cements which are quite familiar to most of us is due to the hydration reactions of cement minerals. There are a number of commercial cements with various hydration speeds. Portland cement is the most common and its chemical and mineral composition is listed in Table 3.5. A number of

Table 3.5 Chemical and mineral compositions of Portland cements

Composition		(1) Common cement	(2) Quick-setting cement	(3) Intermediate-heat cement
Chemical comp.	CaO	64.2	65.9	64.0
	SiO_2	22.2	21.3	24.0
	Al_2O_3	5.2	4.7	4.0
	Fe_2O_3	3.1	2.7	4.2
	SO_3	1.9	2.3	1.3
Mineral comp.	C_3S	43.3	60.0	36.7
	C_2S	31.0	15.8	41.2
	C_3A	8.5	7.9	3.5
	C_4AF	9.4	8.2	12.8

Table 3.6 Chemical compositions of clinker minerals

	SiO_2	Al_2O_3	Fe_2O_3	CaO	MgO	Na_2O	K_2O	Mineral comp.
C_3S	24.83	1.24	0.49	72.23	0.98	0.09	0.14	52.31
C_2S	32.50	2.63	1.03	62.83	0.52	0.20	0.30	26.18
C_3A	5.88	27.43	7.81	53.49	1.97	2.27	1.15	7.87
C_4AF	3.61	24.51	22.08	44.50	4.36	0.37	0.57	9.42

so-called clinker minerals are present in clinkers. Typical clinker minerals are $3CaO \cdot SiO_2$ (alite, abbreviated to C_3S), $2CaO \cdot SiO_2$ (belite, C_2S), $3CaO \cdot Al_2O_3$ (C_3A) and $4CaO \cdot Al_2O_3 \cdot Fe_2O_3$ (C_4AF). These minerals form solid solutions with other oxides. Chemical compositions of the clinker minerals are listed in Table 3.6.

As is well known, when mixed with water, cements solidify by themselves. Although the reactions between cements and water are quite complicated, the hydration reactions of cements can be divided into the following three stages:

(1) Incubation period (S_I and S_{II} in Fig. 3.49). Once a cement contacts water, the hydration of C_3A, C_4AF and plaster takes place swiftly. The hydration can be represented by the following chemical reactions:

$$C_3A + 3CaSO_4 + 32H_2O \rightarrow C_3A \cdot 3CaSO_4 \cdot 32H_2O$$
$$C_3AF + 3CaSO_4 + 32H_2O \rightarrow C_3AF \cdot 3CaSO_4 \cdot 32H_2O$$

These hydration products are commonly called *ettringite* and have a needle-shaped morphology. During this period the amount hydrated does

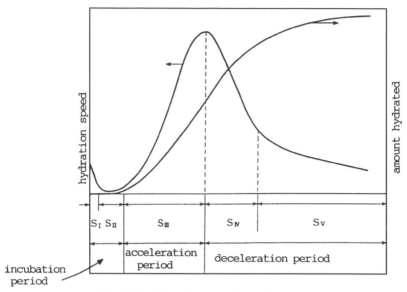

Fig. 3.49 Hydration reaction of a cement

not increase significantly. However, Ca^{2+}, OH^-, K^+ and Na^+ dissolve in water which will be supersaturated quickly with these ions. In addition, scale-shaped hydration products start to precipitate on the surface of C_3S.

(2) Acceleration period (S_{III}). The cement coagulates during this period. The main reaction is the hydration of C_3S which is given by

$$2C_3S + 6H_2O \rightarrow 3CaO \cdot 2SiO_4 \cdot 3H_2O(C_3S_2H_3) + 3Ca(OH)_2$$

The hydration product, $C_3S_2H_3$, has a crystal structure which is very close to that of natural tobermolite and thus is called tobermolite gel.

(3) Deceleration period (S_{IV} and S_V). Because grain boundaries are filled with a large amount of the hydration products, the transport of ions becomes difficult and thus the hydration rate decreases. By the time it reaches S_V, C_3S and C_3A are almost completely hydrated and the hydration of the remaining C_2S takes place, which is the cause of the long-term strength of the cement. The hydration of β-C_2S is as follows:

$$2\beta - C_2S + 4H_2O \rightarrow C_3S_2H_3 + Ca(OH)_2$$

Again, the hydration product forms a tobermolite gel and fills spaces among cement fillers. In order to control the speed of hydration it is necessary to control the chemical and mineral composition, particle size, the ratio of water to cement and the amount of additives.

3.4.6 Liquid Phase Reactions and the Sol–gel Method

The precipitation reaction is one of the most common liquid phase reactions which form a solid phase and has been utilized extensively for the synthesis of raw powders (see Section 3.6.1). Another example of liquid phase reactions is the polymerization of molecules in solutions, which results in the solidification of a sol to a gel. This reaction is called the sol–gel method and is very important technically (Fig. 3.50).

The state in which fine particles (1–100 nm in diameter) are dispersed in either liquid or gases is called colloid. When a colloid is sufficiently fluid and stable for a long period of time, the colloid is called the sol. Rigid solids formed by the evaporation of a solvent from a sol is called the gel. By manipulating the sol–gel transformation properly, it is possible to form a variety of shapes quickly.

When a weak acidic alcohol solution of HCl and water is added to an ethanol solution of tetraethoxysilane (TEOS) $Si(OC_2H_4)_4$ (liquid at room temperature), TEOS hydrates and forms a stable sol which contains oligomers with a small number of the OH^- group. Thin films of SiO_2 can be formed by heating substrates coated with the sol by the dipping method. By increasing the polymerization of the sol at room temperature to 80°C, the sol can be made more viscous and thus can be made into threads. By heating to 400–800°C,

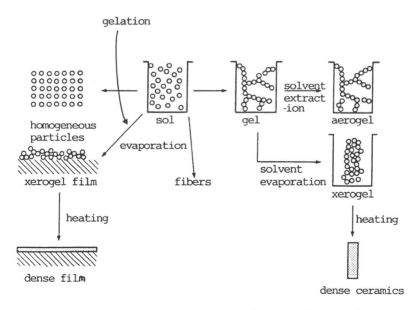

Fig. 3.50 Green forming by the sol–gel method. (From H. Yanagida, *Science of Ceramics*, 2nd edition, Gihohdo, 1993, p. 125: reproduced with permission)

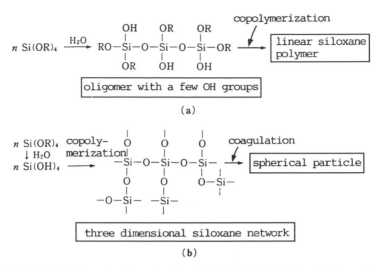

$$n \, Si(OR)_4 \xrightarrow{\text{H}_2\text{O}} \underset{\substack{\underset{\text{OR}}{|} \quad \underset{\text{OH}}{|} \quad \underset{\text{OH}}{|}}{\overset{\substack{\overset{\text{OH}}{|} \quad \overset{\text{OR}}{|} \quad \overset{\text{OR}}{|}}{}}{\text{RO}-\text{Si}-\text{O}-\text{Si}-\text{O}-\text{Si}-\text{OR}}} \xrightarrow{\text{copolymerization}} \boxed{\substack{\text{linear siloxane} \\ \text{polymer}}}$$

$$\boxed{\text{oligomer with a few OH groups}}$$

(a)

(b)

$$\boxed{\text{three dimensional siloxane network}}$$

Fig. 3.51 Hydrolysis and polymerization of $Si(OR)_4$. (a) Acidic solution with a small amount of H_2O; (b) basic solution

SiO_2 fibers can be obtained from the threads. Furthermore, a silica glass can be obtained by heating a completely gelified bulk body to 800°C.

When TEOS is hydrated with a basic solution, the reaction proceeds almost completely to $Si(OH)_4$ which can be further polymerized to form a three-

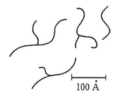 

prior to gelation   near gelation point   gelified: cross bonding
                     intermingled linear chains   at intersections

(a)

prior to gelation   near gelation point   gelified: multiple chai
                     growth and branching   cluster

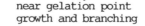

(b)

Fig. 3.52 Morphology of siloxane polymers formed by the sol–gel reaction of TEOS. (a) Acidic condition, addition of a small amount of H_2O; (b) basic condition. (From H. Yanagida, *Science of Ceramics*, Gihohdo, 1993, p. 130: reproduced with permission)

dimensional network of the polymer as indicated in Fig. 3.51(b). When TEOS is hydrated with a large amount of an acidic solution, it tends to form a polymer network rather than polymer chains. Since sol–gel reactions are affected by the reaction temperature, raw materials, the amount of water and pH, microscopic processes of the sol–gel transformations are also affected as indicated in Fig. 3.52.

Various oxide ceramics, glasses and composites are made by the sol–gel method. Metal alkoxides are raw materials for many of these ceramics and some are listed in Table 3.7. A combination of these metal alkoxides are used to produce complex oxide ceramics such as superconductive oxide $YBa_2Cu_3O_{7-x}$. When raw alkoxides are either too expensive or not available, it is possible to use inorganic salts such as halides and nitrates.

Table 3.7 Metal alkoxides used for the sol–gel method

		Alkoxide
Single cation alkoxides		
I A (1) group	Li, Na	$LiOCH_3$ (s), $NaOCH_3$ (s)
I B (11) grp.	Cu	$Cu(OCH_3)_2$ (s)
II A (2) grp.	Ca, Sr, Ba	$Ca(OCH_3)_2$ (s) $Sr(OC_2H_5)_2$, $Ba(OC_2H_5)_2$ (s)
II B (12) grp.	Zn	$Zn(OC_2H_5)_2$ (s)
III A (3) grp.	B, Al, Ga	$B(OCH_3)_3$ (l) $Al(i\text{-}OC_3H_7)_3$ (s) $Ga(OC_2H_5)_3$ (s)
III B (13) grp.	Y	$Y(OC_4H_9)_3$
IV A (4) grp.	Si, Ge	$Si(OC_2H_5)_4$ (l) $Ge(OC_2H_5)_4$ (l)
IV B (14) grp.	Pb	$Pb(OC_4H_9)_4$ (s)
V A (5) grp.	P, Sb	$P(OCH_3)_3$ (l) $Sb(OC_2H_5)_3$ (l)
V B (15) grp.	V, Ta	$VO(OC_2H_5)_3$ (l) $Ta(OC_3H_7)_5$ (l)
VI B (16) grp.	W	$W(OC_2H_5)_6$ (s)
lanthanide	La, Nd	$La(OC_3H_7)_3$ (s) $Nd(OC_2H_5)_3$ (s)
Alkoxides with various alkoxil groups		
	Si	$Si(OCH_3)_4$ (l) $Si(OC_2H_5)_4$ (l) $Si(i\text{-}OC_3H_7)_4$ (l) $Si(t\text{-}OC_4H_9)_4$
	Ti	$Ti(OCH_3)_4$ (s) $Ti(OC_2H_5)_4$ (l) $Ti(i\text{-}OC_3H_7)_4$ (l) $Ti(OC_4H_9)_4$ (l)
	Zr	$Zr(OCH_3)_4$ (s) $Zr(OC_2H_5)_4$ (s) $Zr(OC_3H_7)_4$ (s) $Zr(OC_4H_9)_4$ (s)
	Al	$Al(OCH_3)_3$ (s) $Al(OC_2H_5)_3$ (s) $Al(i\text{-}OC_3H_7)_3$ (s) $Al(OC_4H_9)_3$ (s)
Double cation alkoxides		
	La–Al	$La[Al(i\text{-}OC_3H_7)_4]_3$
	Mg–Al	$Mg[Al(i\text{-}OC_3H_7)_4]_2$, $Mg[Al(s\text{-}OC_4H_9)_4]_2$
	Ni–Al	$Ni[Al(i\text{-}OC_3H_7)_4]_2$
	Zr–Al	$(C_3H_7O)_2Zr[Al(OC_3H_7)_4]_2$
	Ba–Zr	$Ba[Zr_2(OC_2H_5)_9]_2$

grp.: group; l: liquid; s: solid.
S. Sakka, *Science of Sol–gel Method*, Agne Shofu Publishing (1989). Reproduced with permission.

3.4.7 Production of Glasses

There are a number of glasses with various chemical compositions as listed in Table 2.7. Chalcogenide glasses can be made by replacing oxide ions in oxide glasses with Se and Te. Depending on properties, structures and shapes, there is a great variety of glasses such as plate, tempered, crystallized, electrically conductive, optical, IR reflection, photochromic and laser. In this section the manufacturing processes of plate glasses, which have one of the largest production volumes in ceramic science, will be discussed as an example of melt–solidification reactions.

The four major processing steps of glass production are melting, forming, annealing and machining. As indicated in Fig. 3.53, the production of plate glasses also follows similar processing steps. Unlike other ceramics, it is necessary to pay significant attention to the annealing step in the production of glasses. As discussed in Section 2.4, glasses are formed from melts without crystallization. Since glasses are poor thermal conductors, surfaces cool much faster than bulks. When temperatures fall below glass transition temperatures, the shrinkage rate of glasses decreases markedly. On the other hand, bulk glasses whose temperatures are higher than glass transition temperatures have high shrinkage rates. As a consequence, the surfaces are under compressive stresses and the bulks under tensile stresses. Thus in order to reduce the resulting strains, it is necessary to anneal glasses.

The raw materials of common plate glasses are soda ash (Na_2CO_3), calcite ($CaCO_3$), dolomite ($CaCO_3 \cdot MgCO_3$), feldspar and scrap glasses. These materials are ground, mixed together and melted in a tank. Plate glasses are pulled from the melt by various processes which will be discussed in the following section.

3.4.7.1 COLBURN'S PROCESS

As indicated in Fig. 3.54, plate glasses with a constant thickness and width are pulled from the melt between a pair of water-cooled rollers. Glass plates,

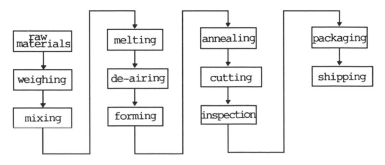

Fig. 3.53 Process flow for the manufacture of plate glasses

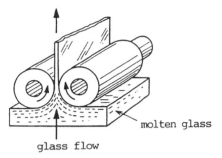

Fig. 3.54   Manufacture of plate glasses by the Colburn process

2–8 mm thick and about 4 m wide, are pulled vertically about 60 cm before they are bent horizontally by heating. The glass plates are then annealed in an annealing oven.

3.4.7.2   FOURCAULT'S PROCESS

A refractory with a center slit, which is called the devitose, is placed on the molten glass and the molten glass is pulled through the slit. The principle of this process is shown in Fig. 3.55. The glass plates are pulled vertically and cut after they are passed through multiple rollers, about 6 m high without bending. Plate glasses, 1–10 mm thick and 1.5–2.5 m wide, can be produced by this process. With an increasing number of devitoses, the amount of production also increases. Commercial production tanks usually hold 3–12 devitoses.

3.4.7.3   PITTBURGH'S PROCESS

This process is very similar to Fourcault's process, but employs a refractory called a draw-bar which is immersed in the molten glass instead of a devitose. Since plate glasses are pulled from a melt such that the free surface of the molten glass becomes the surface of the plate glasses, it is possible to produce plate glasses with flatter surfaces.

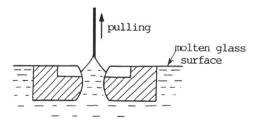

Fig. 3.55   Manufacture of plate glasses by the Fourcault process

3.4.7.4   THE FLOAT GLASS PROCESS

Unlike the conventional pulling processes, in this process plate glasses are produced by floating molten glasses on top of a molten metal with a higher specific gravity. The process was invented by the Pilkington Company in England in 1959. The apparatus is shown schematically in Fig. 3.56.

Molten tin has been selected for the float bath and the molten metal must have the following characteristics:

(1)   It must have a melting point which is lower than the range of temperature in which glasses are molten.
(2)   It must have a higher specific gravity than the molten glass.
(3)   It must have low vapor pressure.
(4)   It must not react with the molten glass.

Plate glasses produced by the conventional pulling processes have to be ground and polished to improve their surface flatness. However, it is not necessary to grind and polish plate glasses produced by this process. In addition, it is also possible to produce large glass plates. The float bath process has been adapted by some Japanese glass companies and it is expected that it will replace all conventional pulling processes in the future.

3.5   SINTERING

3.5.1   The Sintering Phenomenon

When a powder compact is heated at an elevated temperature which is below its melting point, powder particles fuse together, voids between the particles decrease and eventually a dense solid body can be obtained. This phenomenon is called sintering. Sintering processes have been used extensively for the manufacture of porcelain and ironware for hundreds of years, and today sintering is still a very important scientific discipline for the manufacture of a wide variety of industrial materials.

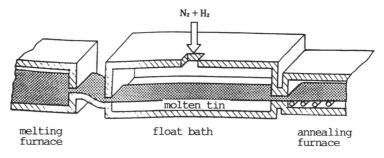

Fig. 3.56   Manufacture of plate glasses by the float glass process

The sum of the surface-free energy of a powder compact is not at its minimum and thus not in equilibrium. When heated, the system tries to decrease its surface-free energy by decreasing its total surface area and thus powder particles are forced to join together. The excess surface-free energy of the powder compact is the driving force of sintering.

The development of microstructures during sintering is quite complicated, but can be distinguished in three stages. At the onset of sintering powder particles fuse together and the area of contact increases gradually, which is called neck growth. In this stage the relative density, which is the density of the powder compact divided by its theoretical density, is about 0.5–0.6 and the shrinkage is about 4–5%.

As sintering proceeds, channel-shaped voids decrease. The relative density increases from 0.6 to 0.95 and the shrinkage also increases from 5% to 20%. This is called the intermediate stage and, in general, particle size increases markedly. When the relative density increases above 0.95, voids remain only at triple points of grain boundaries or inside the grain matrix. Those pores which are open are called open pores and those that are not are called closed. In the final stage sintering proceeds as these pores are gradually eliminated and the relative density increases further. Sintering processes are shown schematically in Fig. 3.57.

3.5.2   The Driving Force of Sintering and Mass Transport

As discussed earlier, the driving force of sintering is the excess surface-free energy of a powder compact. In the following the driving force for curved surfaces will be discussed thermodynamically.

Let γ be the surface-free energy of a curved surface as shown in Fig. 3.58. A perpendicular pressure ΔP is generated at the curved surface due to the radius of the curvature, which is given by

$$\Delta P = 2\{\gamma dl \sin(\theta/2) - \gamma dl \sin(\theta'/2)\}/(dl)^2 \approx \gamma(1/x - 1/\rho) \approx -\gamma/\rho \ (x \gg \rho)$$

$$(3.42)$$

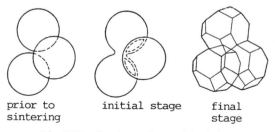

prior to   initial stage   final
sintering   stage

Fig **3.57**   A schematic model of sintering

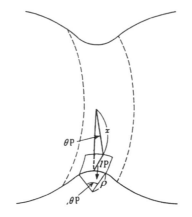

Fig. 3.58 Driving force for sintering on a curved surface

Thus a curved surface has an excess free energy due to ΔP. The excess free energy per mol of the system, ΔF, can be given by

$$\Delta F = \Delta P V \tag{3.43}$$

Since ΔP is negative for concave surfaces, ΔF is also negative. On the other hand, ΔP is positive for convex surfaces and thus ΔF is also positive. As a consequence, mass transport occurs from a convex surface with high surface energy to a concave surface with low surface energy.

Consider the vapor pressure of a curved surface. Let the vapor pressures of flat and curved surfaces be P_0 and P, respectively. The difference in the vapor pressures can be derived by Kelvin's equation,

$$\Delta P = P_0 - P = (M\gamma/\rho dRT)P_0 \tag{3.44}$$

where d and M are the density and the molecular weight, respectively. As is clear from the above equation, the vapor pressure of a convex surface is higher than that of a flat surface and the vapor pressure of a concave surface is lower. As a consequence, a substance vaporizes at a convex surface and condenses at a concave one. This difference in vapor pressure is the driving force of sintering for the evaporation–condensation mechanism.

Next, consider the vacancy concentration n at a curved surface. Because of the perpendicular pressure ΔP acting on the curved surface, the energy required to form a vacancy with a volume of Ω differs $\Delta P \Omega$ compared with a flat surface. Thus the vacancy concentration n at a curved surface can be given by

$$n = n_0 \exp(-\Delta P\Omega/RT) \tag{3.45}$$

where n_0 is the vacancy concentration at a flat surface. From Eqs (3.42) and (3.45) and assuming that $\Omega\gamma/\rho kT \ll 1$,

$$\Delta n = n - n_0 \approx n_0 \Omega \gamma / \rho k T \qquad (3.46)$$

This equation indicates that the vacancy concentration on a convex surface is smaller than that on a flat surface and the vacancy concentration on a concave surface is higher. As a consequence, vacancies flow from a neck to a convex area and thus a mass flow occurs in the reverse direction. The difference in vacancy concentration is the driving force of sintering due to the diffusion mechanism.

When there is a liquid phase around a particle, there exists a difference in solubility between flat and curved surfaces. As in Eq. (3.46) the difference, Δs, can be given by

$$\Delta s = s - s_0 \approx (M\gamma / \rho d R T) s_0 \qquad (3.47)$$

The solubility of a convex surface is higher and that of a concave one is lower. This difference in solubility is the driving force of liquid phase sintering.

3.5.3 Sintering Mechanisms and their Rate Equations

Depending on the mode of mass transport as indicated in Fig. 3.59, diffusion mechanisms can be classified into the following four major categories: evaporation–condensation, diffusion, dissolution–precipitation, and flow. The predominant diffusion mechanism for a given system can be identified from the analysis of rate equations for neck growth and shrinkage during the initial stage of sintering. In real situations sintering processes are a combination of these diffusion mechanisms (Table 3.8).

3.5.3.1   THE EVAPORATION AND CONDENSATION MECHANISM

The difference in vapor pressure for a sphere-to-sphere contact shown in Fig. 3.59 can be given by Eq. (3.44):

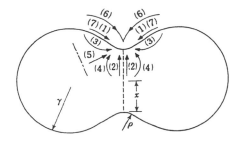

Fig. 3.59 Mass transport paths for sintering. (1) S→neck SD: (2) GB→neck GBD; (3) S→neck VD; (4) GB→neck VD; (5) DL→neck VD; (6) S→neck EC; (7) S→neck DP; S—surface GB, grain boundary; DL—dislocations in grain matrix; SD—surface diffusion; GBD—grain boundary diffusion; VD—volume diffusion; EC—evaporation–condensation; DP—dissolution–precipitation

Table 3.8   Diffusion mechanisms and their rate equations

Mechanism	Neck growth eq.	Shrinkage rate eq.	Source	Sink
Volume diffusion (I)	$\dfrac{x^5}{r^2} = \left(\dfrac{40\gamma\Omega D}{kT}\right)t$	$\left(\dfrac{\Delta L}{L_0}\right)^{5/2} = \left(\dfrac{20\gamma\Omega D_r}{\sqrt{2}r^3 kT}\right)t$	Neck	Grain boundary
Volume diffusion (II)	$x^3 = \left(\dfrac{Q\gamma\Omega D_r}{kT}\right)t$	$\left(\dfrac{\Delta L}{L_0}\right)^{3/2} = \left(\dfrac{B\gamma\Omega D_r}{kT}\right)t$	Dislocation	Grain boundary
Volume diffusion (III)	$\dfrac{x^5}{r^2} = \left(\dfrac{40\gamma\Omega D_r}{kT}\right)t$	$\left(\dfrac{\Delta L}{L_0}\right) = 0$	Neck	Convex region
Grain boundary diffusion	$\dfrac{x^6}{r^2} = \left(\dfrac{96\gamma\Omega D_g}{kT}\right)t$	$\left(\dfrac{\Delta L}{L_0}\right)^3 = \left(\dfrac{3\gamma\Omega D_g}{r^4 kT}\right)t$	Neck	Grain boundary
Surface diffusion	$\dfrac{x^7}{r^3} = \left(\dfrac{56\gamma\Omega D_s}{kT}\right)t$	$\left(\dfrac{\Delta L}{L_0}\right) = 0$	Neck	Convex region

Notes:
D_r: volume diffusion coefficient, D_g: grain boundary diffusion coefficient, D_s: surface diffusion coefficient, L_0: dimension of green body, ΔL: sintering shrinkage.

$$\Delta P = P_2 - P_1 = (2M\gamma/\rho dRT)P_0 \tag{3.48}$$

This difference in vapor pressure is the driving force of sintering and the growth of a neck proceeds as a consequence. Assuming that Langmuir's equation is valid for mass transport in a gas phase, the flow rate, m, of a substance to the neck area can be given by

$$m = \alpha\Delta P(M/2\pi RT)^{1/2} \tag{3.49}$$

where α is the geometric constant. Let the radius of the sphere be r, the volume of the neck area V, the area A and the radius of curvature ρ. Then the following relationships can be derived: $P = x^2/2r$, $A = \pi^2 x^3/r$ and $V = \pi x^4/2r$. The condensation rate of the substance at the neck is given by

$$dV/dt = (m/d)A \tag{3.50}$$

The solution of Eq. (3.50) with the relationships derived above gives

$$x^3/r = \{3\alpha\gamma M^{3/2}P_0\pi^{1/2}/[(2)^{1/2}d^2 R^{3/2}T^{3/2}]\}t \tag{3.51}$$

NaCl is considered to sinter by the evaporation–condensation mechanism.

3.5.3.2   THE DIFFUSION MECHANISM

As indicated in Fig. 3.59, the diffusion mechanism can be subclassified into the following three mechanisms; surface, grain boundary and volume. As discussed earlier, the driving force is the difference in vacancy concentration between convex and neck areas. Vacancies are generated either at convex surfaces or at dislocations (mainly edge dislocations) in a grain matrix, and are called vacancy sources. On the other hand, vacancies are eliminated at concave surfaces, grain boundaries and dislocations, and are called vacancy sinks. Thus depending on the combination of source and sink, the diffusion mechanisms can be subdivided further. Rate equations for a sphere-to-sphere contact are listed in Table 3.8.

Except in substances with high vapor pressure such as NaCl, the sintering of powders proceeds by the volume diffusion mechanism.

3.5.3.3 THE FLOW MECHANISM

A perpendicular pressure pointing to the center of a powder particle acts on a convex surface and that pointing away from the center acts on a concave one. When a substance has fluidity, mass transport proceeds under the difference in the perpendicular pressure. Sintering by this type of mass transport is called the flow mechanism. Depending on the mode of mass flow, the sintering mechanism is called either the viscous or the plastic flow mechanism. The rate equation for the viscous flow mechanism is given by

$$x^2/r = k(\gamma/\eta)t \qquad (3.52)$$

$$\Delta L/L_0 = k'(\gamma/\eta r)t \qquad (3.53)$$

where η is the viscosity of the substance and k and k' are the constants. It has been confirmed experimentally that Eq. (3.52) is valid for sintering glass spheres.

3.5.3.4 THE DISSOLUTION–PRECIPITATION MECHANISM

The formation of a liquid phase promotes the sintering of substances which contain multiple solid phases such as cermets and porcelain. In a binary phase system, let T_1 and T_2 be the melting point of Phase I and Phase II, respectively, where $T_1 < T_2$. When the sintering temperature T is between these two melting points ($T_1 < T < T_2$), the phase with a low melting point melts and forms a liquid phase. When the liquid phase promotes sintering, this is called liquid phase sintering, which can be distinguished from solid phase sintering discussed above.

A new and unique phenomenon can be observed in liquid phase sintering. In the initial phase of sintering the presence of a liquid phase which wets the solid phase allows the rearrangement of particle packing by gliding. Subsequently the substance dissolved from convex surfaces is transported to concave surfaces and precipitates out, which is the densification stage of dissolution and precipitation sintering. When either dissolution or precipitation is rate-controlling, the rate equation is given by

$$\Delta L/L_0 = k_1 r^{-4/3} t^{1/3} \qquad (3.54)$$

When the diffusion of a substance in the liquid phase is rate-controlling, the rate equation is given by

$$\Delta L/L_0 = k_2 r^{-1} t^{1/2} \tag{3.55}$$

As sintering proceeds, growth in neck areas and an increase in grain size can be observed. The presence of liquid phase sintering has been confirmed in Fe–Cu and Si_3N_4–Y_2O_3 systems.

3.5.4 Pressureless Sintering

The sintering of powder compacts can be classified into two major categories. One is called pressureless sintering, which involves the forming of desired shapes prior to sintering. The other category is pressure sintering, i.e. the densification of powder compacts under pressure during sintering. In this section the forming methods employed for pressureless sintering will be discussed. A number of innovations have been introduced to promote densification by pressureless sintering. In addition, sintering conditions varies from substance to substance. In order to shorten our discussion on pressureless sintering, sintering of individual substances will, regretfully, be omitted.

3.5.4.1 THE PRESSURE FORMING METHOD

When powder is compacted in a metallic die made of steel, etc. and pressed by plungers and punches, various shapes such as pellets, rectangles and plates can be formed. In general, powder is pressed uniaxially using a die shown in Fig. 3.60. During pressing, the powder compacts experience various pressure distributions depending on the magnitude of applied pressure, which tend to cause formation of fissures and cracks in sintered bodies. Thus care is necessary during pressing. When the formability of the powder alone is not sufficient, it is necessary to evacuate during pressing as well as to add forming additives such as water, cellulose and alcohol.

The rubber press method may be employed to eliminate density variation in powder compacts due to pressure distribution during uniaxial pressing. Powder is compacted in rubber bags and immersed in water or oil. When these media are pressurized, the powder compacts form desired shapes uniformly under isostatic pressure.

3.5.4.2 THE SLIP-CASTING METHOD

Slurries with proper viscosity are prepared by dispersing powder in a solvent. Subsequently, the slurries are cast in molds which have cavities with desired shapes. There are two types of slip-casting, namely drain casting and care casting (or solid casting). In the former the molds are first filled with slurries.

force

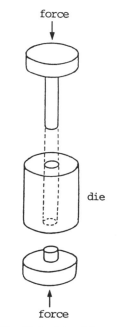

die

force

Fig. 3.60 Uniaxial cold pressing die

After a time, the slurries are drained from the molds. This type of casting is suitable for obtaining shapes with hollow centers. In the latter casting, the entire amount of slurries which fills the mold is used to form a dense finished shape. The two types of casting are shown schematically in Fig. 3.61.

The doctor blade method is a derivative of slip casting. As shown in Fig. 3.62, a batch of slurry is poured on a glass plate or plastic film and its thickness is controlled by the doctor blade. Alumina substrates for IC applications are produced by this method.

3.5.4.3   THE EXTRUSION METHOD

A dough with much higher viscosity than that of slurries is prepared by mixing a powder with a solvent and is subsequently extruded through an opening to shape green forms. The extrusion method is shown schematically in Fig. 3.63. Products with relatively simple geometries can be produced by this method (for example, electrical insulators, combustion tubes and carbon electrodes).

3.5.4.4   THE INJECTION MOLDING METHOD

A powder is mixed with a thermoplastic resin and is heated before the mixture is injected into a mold with a desired shape. The thermoplastic resin gives the

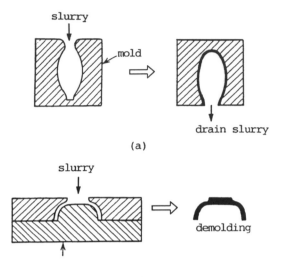

(a)

(b)

Fig. 3.61 The slip-casting method. (a) Drain casting; (b) solid casting

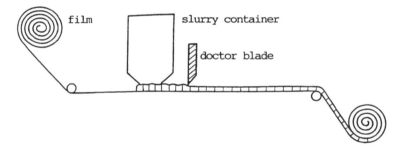

Fig. 3.62 The doctor blade method

necessary plasticity during injection molding and can be removed easily from the green forms by thermal decomposition prior to sintering. It is possible to green form fairly complicated shapes such as turbochargers by this method.

3.5.5 Pressure Sintering

Pressure sintering combines two steps of pressureless sintering, namely forming and sintering, into one. In pressure sintering, a powder is compacted in a die and sintered at elevated temperatures under pressure. Compared to pressureless sintering, pressure sintering is more expensive, not amenable to producing complicated shapes and not suited for mass production. On the

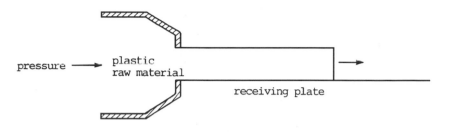

Fig. 3.63   The extrusion method

other hand, pressure sintering lowers the sintering temperature, makes it possible to produce sintered bodies with near-theoretical density for a relatively short period of time, a sintered body which is obtained only under high pressure and sintered bodies of nitrides and carbides which are very difficult to sinter without pressure.

The most common pressure sintering method is called hot pressing. A hot pressing apparatus is shown in Fig. 3.64. A powder is compacted in a die made of graphite, Al$_2$O$_3$ or SiC and pressed uniaxially while heated to high temperatures. The die is heated by a resistance heater, by passing an electric current through the die itself if the die is made of an electrically conductive material or by RF induction heating. The maximum pressure exerted on the powder compact depends on the selection of die material or on the mode of pressing. Die materials for uniaxial hot pressing are listed in Table 3.9.

Selection of a die material is a major challenge in hot pressing. Hot isostatic pressing (HIP) is one alternative to hot pressing. HIP is a derivative of the

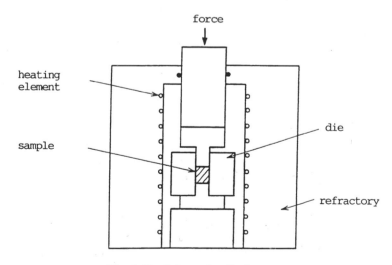

Fig. 3.64   Schematic of a hot press

Table 3.9 Die materials for uniaxial hot pressing

Material	Max. temp. (°C)	Max. press (bar)	Material	Max. temp (°C)	Max. press (bar)
Graphite	2500	700	W	1500	2450
Alumina	1200	2100	Mo	1100	210
Zirconia	1180	1050	Inconel X	1100	
SiC	1500	2800	Stainless steel	1100	
TaC	1700	560			
WC, TiC	1400	1700			
TiB$_2$	1200	1050			

T. Vasilos and R. M. Spriggs, *Prog. Ceram. Sci.*, **4**, 96 (1966).

rubber press method and employs inert gas such as Ar as a pressurization medium and sinters in a high-temperature, high-pressure vessel. Powders are compacted in containers made of Fe, Pt and Ni. WC, alumina and MgO with a theoretical density of 99.8–99.9% can be produced by HIPping them under 1000–2000 bar. It is also common to HIP high-density sintered bodies after pressureless sintering.

It is known that Si$_3$N$_4$ can be sintered with few sintering aids when HIPped under a nitrogen gas pressure of 10–50 atm. Since high-pressure nitrogen can suppress the following decomposition reaction

$$Si_3N_4 \rightarrow 3Si + 2N_2$$

it is possible to raise the sintering temperature and thus promote densification.

Diamond and cubic BN which are vital for the manufacture of super-hard cutting tools are obtained only at high temperatures and high pressures as indicated in Fig. 3.65. The sintering of these materials requires higher pressures (20–50 kbar) which can be generated by tetrahedral or octahedral anvil machines. The tips of the anvils are made of diamond.

Examples of sintered bodies produced by pressure sintering are listed in Table 3.10.

3.5.6 Manufacture of Porcelain

Porcelain is a very important class of ceramics which have been used daily in our lives since ancient times. As indicated in Table 3.11, porcelain can be classified into four major groups depending on whether it is porous enough to absorb water and whether it has a glaze coating. Porcelain can be further subclassified depending on its composition, microstructure and manufacturing processes. In this section manufacturing processes of a typical porcelain will be discussed.

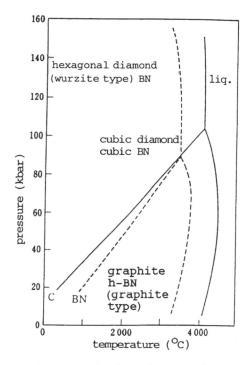

Fig. 3.65 Temperature–pressure phase diagrams of carbon and boron nitride

Table 3.10 Examples of pressure sintered materials

Oxides	MgO, Al_2O_3, ZrO_2, $BaTiO_3$, etc.
Nitrides	Si_3N_4, TiN, cubic BN, sialon, etc.
Carbides	SiC, WC, TiC, diamond, B_4C, etc.
Borides	LaB_6, TiB_2, ZrB_2, NbB_2, etc.

Raw materials of porcelain are siliceous stone whose main component is SiO_2, clay which contains kaolinite $Al_2(Si_2O_5)$ $(OH)_4$ and mica, and feldspar which contains K^+, Na^+ and Ca^{2+}. The functions of these ingredients are as follows: the siliceous stone forms a skeletal frame of porcelain, the clay controls plasticity of green mixtures and the feldspar promotes the formation of glassy phases.

The process flow for the manufacture of porcelain is shown in Fig. 3.66. After drying, green porcelain is fired in a variety of ovens. In general, porcelain is fired three times, i.e. primary, main and paint firings. Primary firing is always conducted in air. While some sintering takes place during primary firing, physically and chemically adsorbed water is eliminated almost completely from

Table 3.11 Classification of porcelain

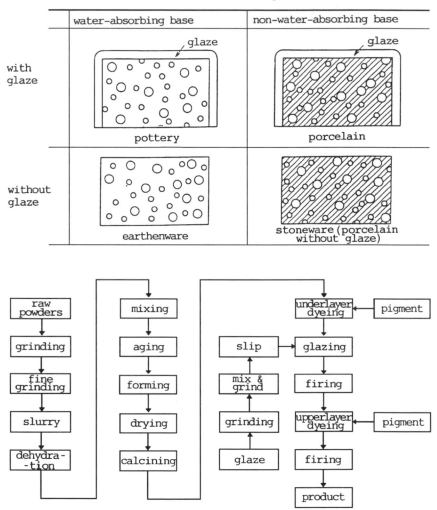

Fig. 3.66 Process flow for the manufacture of porcelain

porcelain bodies by the firing. The main firing can be conducted either in an oxidizing environment or in a variable atmosphere from oxidizing to neutral to reducing for white porcelain ware. The atmosphere is controlled to eliminate the coloration of porcelain by colorizable components such as Fe. Thermal decomposition of carbonates and organic substances, vitrification, and melting of a glaze take place during the main firing. After the main firing, the porcelain bodies are painted and fired again to complete the manufacturing process.

3.6   OTHER CERAMIC PROCESSES

3.6.1   Powder Processes

There is a wide variety of methods to produce ceramic powders. The characteristics of ceramic powders influence significantly their processability as well as the nature of the final ceramic products. New technologies have been developed to control grain sizes and their distribution as well as morphologies. In the following, various methods to produce ceramic powders will be discussed briefly.

3.6.1.1   THE GAS PHASE METHOD

This is the so-called chemical vapor deposition method (CVD). Fine particles of oxides, nitrides and carbides are produced by the reaction of gaseous metal salts, which have high vapor pressures, with various gases. Table 3.12 lists examples of the gas phase reactions and Fig. 3.67 shows the dependence of formation-free energies of various gas phase reactions listed in Table 3.12 on temperature. The formation of solid phases by the gas phase reactions involves two stages, namely nucleation and growth.

In order to produce fine powders, it is necessary to have an adequate amount of supersaturation during homogeneous nucleation. When supersaturation is small, crystal growth overwhelms nucleation and the formation of large particles or a single crystal results. The equilibrium vapor pressure of a solid is constant at a constant temperature. Thus in order to increase the degree of supersaturation, it is necessary to increase the vapor pressure of the solid which is formed by the gas phase reaction. By improving supersaturation, it is possible to produce high-purity, ultra-fine particles with particle sizes less than

Table 3.12   Examples of gas phase reactions

1	$SiCl_4 + O_2 \rightarrow SiO_2 + 2Cl_2$
2	$TiCl_4 + O_2 \rightarrow TiO_2(A) + 2Cl_2$
3	$TiCl_4 + O_2 \rightarrow TiO_2(R) + 2Cl_2$
4	$TiCl_4 + 2H_2O \rightarrow TiO_2(A) + 4HCl$
5	$AlBr_3 + 3/4O_2 \rightarrow 1/2Al_2O_3 + 3/2Br_2$
6	$AlCl_3 + 3/4O_2 \rightarrow 1/2Al_2O_3 + 3/2Cl_2$
7	$FeCl_3 + 3/4O_2 \rightarrow 1/2Fe_2O_3 + 3/2Cl_2$
8	$ZrCl_4 + O_2 \rightarrow ZrO_2 + 2Cl_2$
9	$SiCl_4 + 2H_2 + 2/3N_2 \rightarrow 1/3Si_3N_4 + 4HCl$
10	$SiCl_4 + 4/3NH_3 \rightarrow 1/3Si_3N_4 + 4HCl$
11	$TiCl_4 + 1/2N_2 + 2H_2 \rightarrow TiN + 4HCl$
12	$TiCl_4 + NH_3 + 1/2H_2 \rightarrow TiN + 4HCl$
13	$VCl_4 + NH_3 + 1/2H_2 \rightarrow VN + 4HCl$
14	$TiCl_4 + CH_4 \rightarrow TiC + 4HCl$

Note: The numbers correspond to those in Fig. 3.67.

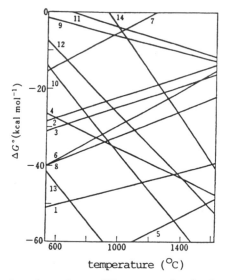

Fig. 3.67 Temperature dependence of free energy for the formation of oxides, nitrides and carbides by the gas phase reaction. (From Japan Chemical Society (ed.), *Treatise of Chemistry IX: Inorganic Reactions with Solids*, Society Publications Center, 1975, p. 146: reproduced with permission)

1 μm. In general, by the gas phase method it is possible to obtain powders with uniform particle sizes and good dispersability.

In addition, it is also possible to oxidize or nitride metallic vapors directly. For example, when the vapor of a zinc metal is oxidized in air by spraying the vapor of the molten zinc through a nozzle into air, fine particles of ZnO of about 0.5 μm can be obtained. By controlling processing parameters, it is possible to obtain various sizes of ZnO particles. By this method, powders of MgO and CdO in addition to ZnO have been produced.

3.6.1.2 THE LIQUID PHASE METHODS

The Precipitation Method

When aqueous solutions of various salts react, compounds with low solubility precipitate out of the solutions. Ceramic powders can be obtained after washing, drying and calcining the precipitates. By employing various salts and controlling processing parameters such as temperature, it is possible to control the particle size of ceramic powders produced. This method is especially suited for the production of fine particles. Examples of the precipitation reactions employed for the production of ceramic powders are:

$$CaCl_2 + Na_2CO_3 \rightarrow CaCO_3 + 2NaCl$$
$$Fe(NO_3)_3 + 3NaOH \rightarrow Fe(OH)_3 + 3NaNO_3$$
$$Pb(NO_3)_2 + (NH_4)_2S \rightarrow PbS + 2NH_4NO_3$$

By the precipitation method, it is also possible to obtain ceramic powders with mixed cations and of solid solutions in addition to ceramic powders with a single cation. For example, $MgO \cdot Al_2O_3$ spinel can be produced as follows. Aqueous solutions of $Mg(NO_3)_2$ and $Al(NO_3)_3$ are mixed at an equimolar ratio. A hydroxide is formed by adding an ammonia solution and precipitates out of the solution. A homogeneous powder of spinel can be obtained by calcining the hydroxide precipitate. When two or more salts are precipitated together, the method is called co-precipitation.

Hydrolysis of metal alkoxides has also been used extensively to produce ceramic powders with desired particle size and morphology. The hydrolysis of titanium tetraisopropoxide can be represented by

$$Ti(OC_4H_9)_4 + 2H_2O \rightarrow TiO_2 + 4C_4H_9OH$$

This reaction is a type of polymerization reaction and can be classified as a derivative of the sol–gel reaction.

The Spray Drying and Spray Decomposition Methods

When droplets produced by the atomization of an aqueous solution are dried, crystallites thus produced tend to be smaller than the droplets. When the drying speed is high, it is possible to produce very fine powders. This method is called spray drying and has been used to produce ferrite powders.

When an aqueous solution is atomized at high temperatures, the salt in the solution is decomposed thermally and thus spherical particles with a uniform particle size can be produced. This method is called spray decomposition.

The Freeze-drying Method

As indicated in Fig. 3.68, an aqueous solution of a metal salt at a condition A is cooled to another condition B where the salt and ice co-exist. When the mixture is evacuated at a low pressure, crystals of the salt can be obtained after the sublimation of ice. By quenching from A to B, it is possible to make fine powders. The powder method based on this principle is called freeze drying and has been used extensively for the production of biochemical materials such as proteins as well as for the production of fine powders of WC and metals.

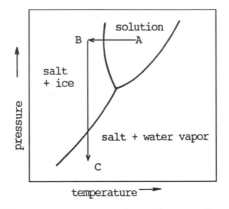

Fig. 3.68   Temperature–pressure phase diagram of an aqueous solution

3.6.1.3   THE SOLID PHASE METHOD

The Thermal Decomposition Method

Ceramic powders can be produced by the thermal decomposition of carbonates, hydroxides and oxalates. Since the ceramic powders produced by this method tend to have morphologies which are very close to those of mother salts, it is very important to choose mother salts carefully. Examples of ceramic powders produced by this method are MgO by the dehydration of $Mg(OH)_2$ and NiO by the decarborization of $NiCO_3$. $BaTiO_3$ powder by the thermal decomposition of titanium barium oxalate is an example of ceramic powders with double cations produced by this method. The $BaTiO_3$ powder has a BaO/TiO_2 ratio which is very close to unity compared to the powder produced by the reaction of $BaCO_3$ with TiO_2.

The solid Phase Reaction Method

Ceramic powders can be produced by mixing two or more powders and by heating them at temperatures lower than the lowest melting point of the mother powders. A variety of complex oxide powders such as spinel, barium titanate and ferrites have been produced by this method. Since two or more powders react to produce complex oxides, it is very important to mix the mother powders very uniformly.

3.6.2   Thin-film Processes

As listed in Table 3.13, there exists a wide variety of methods to produce ceramic thin films. Each method has its unique characteristics and plays a crucial role in determining the characteristics of thin films produced. Thus it is important not only carefully to choose deposition conditions, such as

Table 3.13 Classification of thin-film methods

Gas phase method	Chemical reaction method	┌CVD method ├Chemical transport method ├Substrate reaction method └Spray pyrolysis method
	Physical vaporization	┌Vacuum evaporation method ├Sputtering method ├Ion plating method └Plasma spray method
Liquid phase method		┌Sol–gel method ├Solution epitaxy method └Melt epitaxy method
Solid phase method		┌Coat-decomposition method └Precipitation method

temperature, pressure, atmosphere and starting materials, but also to select the deposition method carefully.

3.6.2.1 THE GAS PHASE METHOD

The Chemical Vapor Deposition Method (CVD)

A vaporizable compound which contains an element in a thin film is transported by a carrier gas to a reaction chamber. A solid thin film is formed on a substrate at elevated temperatures by the chemical reaction of the compound. Horizontal reactors (Fig. 3.69) have been used extensively for the deposition of thin films by the CVD method. Depending on the mode of excitation employed, CVD methods are called thermal CVD, optical (laser) CVD or plasma CVD. When a metal organic compound is employed as a starting material, the method is called the metal organic CVD (MOCVD). When single crystals are grown epitaxially by molecular beams from the Knudsen cell, the method is called molecular beam epitaxy (MBE).

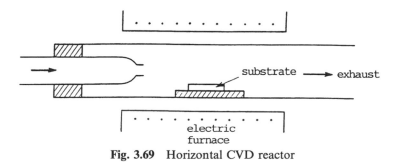

Fig. 3.69 Horizontal CVD reactor

Various chemical reactions employed for CVD are listed in Table 3.14. As can be seen in the table, the vaporizable compounds are mostly halides and hydrocarbons. Since the deposition of thin films is based on heterogeneous nucleation and growth, the morphology and microstructure of a thin film depend strongly on the degree of supersaturation and the rate of chemical reaction.

The Chemical Vapor Transport Method

This method is not as common as the CVD method, but is well suited to the formation of thin films of metal compounds with variable valences. The transport of gaseous species occurs in either open or closed reaction tubes with a thermal gradient. The rate of film growth is small due to the fact that the reaction takes place in a condition very close to equilibrium, but the resulting thin films have good uniformity. It is also possible to grow epitaxial thin films by this method.

The Substrate Reaction Method

Thin films are formed by the direct reaction of substrates with gases. Thin films of oxides, nitrides and carbides are formed directly on metal surfaces. This

Table 3.14 Examples of thin films produced by the CVD method

	Thin film	Raw gases	Carrier gas	Reaction temp. (°C)
Oxide	Al_2O_3	$AlCl_3 + H_2O$	$Ar + O_2$	800–1000
	SiO_2	$SiCl_4 + H_2O$	$Ar + O_2$	800–1100
	Fe_2O_3	$Fe(CO)_5$	$H_2 + CO$	100–300
	ZrO_2	$ZrCl_4$	$H_2 + CO$	800–1000
	$NiFe_2O_4$	$NiBr_2 + FeBr_2 + H_2O$	$Ar + O_2$	750–1100
Nitride	Si_3N_4	$SiCl_4$	$N_2 + H_2$	1000–1600
	BN	BCl_3	$N_2 + H_2$	1200–1500
	TiN	$TiCl_4 + NH_3$	H_2	1100–1700
	AlN	$AlCl_3$	$N_2 + H_2$	1200–1600
	ZrN	$ZrCl_4$	$N_2 + H_2$	2000–2700
Carbide	SiC	CH_3SiCl_3	Ar	1400
	TiC	$TiCl_4 + CH_3$	H_2	1300–1700
	WC	$WCl_6 + C_6H_5CH_3$	H_2	1000–1500
	BeC	$BeCl_3 + C_6H_5CH_3$	H_2	1300–1400
Boride	AlB	$AlCl_3 + BCl_3$	H_2	1000–1300
	ZrB_2	$ZrCl_4 + BBr_3$	H_2	1700–2500
	SiB	$SiCl_4 + BCl_3$	H_2	1100–1300
	TiB_2	$TiCl_4 + BBr_3$	H_2	1100–1300

Fundamental of Thin Film Formation, p. 173, T. Asamaki Nikkankogyo Newspaper (1977).

method has been applied commercially for oxidation of silicon for electronic applications, formation of protective oxide films on magnetic iron powders and nitridation of steels. Because of the comparatively low temperatures involved, thin films produced by this method tend to have poor crystallinity. Another disadvantage of this method is the preferential reaction at grain boundaries when polycrystalline substrates are employed.

The Spray Pyrolysis Method

Thin films are formed on heated substrates by spraying aqueous solutions of metal compounds such as halides. For example, a transparent conductive film of SnO_2 on a glass surface is formed by the following spray pyrolysis reaction:

$$SnCl_4 + 2H_2O \rightarrow SnO_2 + 4HCl$$

The spray pyrolysis method is very simple and has also been used for the deposition of ferrite thin films.

The Vacuum Evaporation Method

A substance with a high vapor pressure is heated by a tungsten heater to evaporate. The evaporated substance impinges on substrates and forms thin films (Fig. 3.70). This method is best suited to the formation of metallic thin films. The deposition of oxides and nitrides by this method requires very high temperatures because these compounds have low vapor pressures. Since the compounds are not stable at very high temperatures, the composition of thin films thus formed is not necessarily that of the starting compound. In order to minimize these problems, derivatives of this method have been attempted, i.e. (1) the reaction–vaporization method, by which thin films are formed by the direct reaction of metal vapor in an atmosphere with low oxygen partial

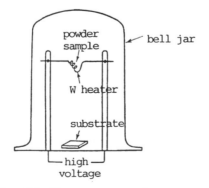

Fig. 3.70 Vacuum evaporation apparatus

pressure and (2) the flash method, by which thin films are formed by the flash evaporation of a small quantity of the starting powder. However, these methods are not common.

The Sputtering Method

When a solid surface is bombarded with ions, neutral atoms, molecules and ions are emitted from the solid surface in addition to electrons, electromagnetic waves and gases. Thin films are grown by capturing these neutral atoms and molecules on substrates. A sputtering apparatus is shown schematically in Fig. 3.71. The vacuum chamber is evacuated and refilled with 10^{-2} to 10^{-4} Torr Ar gas. A plasma is generated by applying a high voltage (200–3000 V) between a target, which is either a metal or a sintered disk, and a substrate. When the target is bombarded with Ar ions, atoms and ions are knocked out of the target and are deposited on the nearby substrate.

Thin films of oxides and nitrides are formed by the reactive sputtering method. In addition to Ar in the vacuum chamber, reactive gases such as O_2 and N_2 are introduced during the sputtering of a metal target. Thin films of complex oxides and solid solutions have been deposited using targets with proper compositions that are determined not only by the composition but also by the sputtering yield of various metals. Examples of ceramic thin films deposited by the sputtering method are listed in Table 3.15.

The slow rate and the marked increase in substrate temperature are two major problems in the deposition of thin films by the conventional DC sputtering method. RF and magnetron sputtering methods have been introduced to overcome these problems.

The Ion Plating Method

This method is a hybrid of vacuum deposition and sputtering. Gaseous species evaporated from an evaporation source are ionized by plasma discharge between a substrate and an evaporation source by applying a high voltage in a

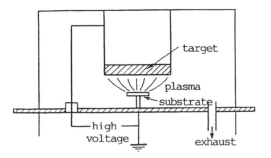

Fig. 3.71 Sputtering apparatus

Table 3.15 Examples of thin films produced by the reactive sputtering method

Thin film	Sputtering method	Anode mat'l	Discharge gas (Torr)		Substrate temp. (°C)
AlN	RF	Al	Ar	4×10^{-3}	250
			N_2	2×10^{-3}	
NbN	Non-symmetric AC	Nb	Ar	3×10^{-2}	600
			N_2	2×10^{-5}	
PtO_2	DC two poles	Pt	O_2	5×10^{-3}	
TaC	DC two poles	Ta	Ar	3×10^{-3}	400
			CH_4	4×10^{-5}	
			or		
			CO	2×10^{-5}	
Ta_2N	DC two poles	Ta	Ar	3×10^{-3}	400
			N_2	4×10^{-5}	
TaN	DC two poles	Ta	Ar	3×10^{-3}	400
			N_2	$1-5 \times 10^{-4}$	
Ta_2O_5	DC two poles	Ta	Ar	3×10^{-3}	700–900
			O_2	9×10^{-3}	
TiN	RF	Ti	Ar	5×10^{-3}	RT
			N_2	4×10^{-3}	
TiO	RF	Ti	Ar	5×10^{-3}	RT
			O_2	4×10^{-3}	

K. Kinbara, *Fundamental Technologies of Thin Films*, p. 61, Tokyo University Press (1979). Reproduced with permission.

vacuum chamber filled with low-pressure Ar. The ions thus formed are accelerated and deposited on the substrate to form thin films. The advantages of this method are (1) it is not necessary to heat substrates, and (2) thin films have good crystallinity because of the high reactivity of ions on the substrate.

With improvements in heating and ion acceleration methods, it is now possible to deposit thin films of oxides, nitrides and carbides by the ion plating method.

The Plasma Spray Method

A raw powder is fed to a plasma flame of either Ar or Ne and the superheated powder is sprayed onto the substrates to form thin films. The advantages of this method are the high rate of film growth and the ability to deposit films in air. Since the substrates are heated substantially, it is necessary to use high-temperature substrates such as ceramics and Pt. Thin films of complex oxides such as various ferrites have been produced commercially by this method.

3.6.2.2 THE LIQUID PHASE METHOD

The Sol–gel Method (see Section 3.4.6)

A sol is prepared from metal alkoxides, and other organic and inorganic salts. Thin films are formed on substrates by drying and heating of a sol which is coated on the substrates by various methods shown in Fig. 3.72. Since thin films can easily be formed on a variety of substrates with widely different sizes and shapes, this method has been used extensively to form thin films of dielectric materials, piezoelectric materials, superconductive materials and ferrites.

The Liquid Phase Epitaxy Method

This method has been used to grow single-crystal films of ceramics from a liquid phase and is a derivative of the flux method for the growth of single crystals. Single-crystal films are grown epitaxially by dipping single-crystal substrates into a supersaturated melt solution. Since it is necessary to employ single-crystal substrates which do not dissolve in the melt, the limited choice of the substrate is a major disadvantage. The growth of $Gd_3Ga_5O_{12}$(GGG) films by this method has been reported.

The Melt Epitaxy Method

As shown in Fig. 3.73, a sample of ceramic powder is melted on a single-crystal substrate and the molten ceramic substance is cooled gradually to form a

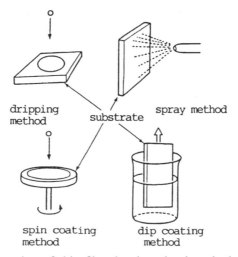

Fig. 3.72 The formation of thin films by the sol–gel method. (From H. Yanagida, *Science of Ceramics*, 2nd edition, Gihohdo, 1993, p. 134: reproduced with permission)

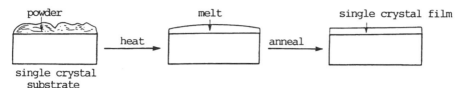

Fig. 3.73   Principle of the melt epitaxy method

single-crystal film epitaxially on the substrate. Single-crystal films of LiNbO$_3$ have been grown on LiTaO$_3$ substrates by this method.

3.6.2.3   THE SOLID PHASE METHOD

The Coat-thermal Decomposition Method

A thin film is formed by the high-temperature thermal decomposition of a metal organic compound which is dissolved in an organic solvent and coated on a substrate. ZnO films are formed from a *n*-butanol solution of Zn(C$_4$H$_7$O)$_2$. Thin films produced by this method include PbO, VO$_2$ and complex oxides such as ferrites, BaTiO$_3$ and oxide superconductors.

The Precipitation Method

A desired substance is precipitated on a substrate as discussed in Section 3.6.1 and is sintered at high temperatures to form a polycrystalline thin film. This method has been applied to form thin films of BaTiO$_3$ and ferrites.

3.6.3   Fiber Processes

Ceramic fibers are produced by a variety of methods. In this section processes to produce glass, optical and ceramic fibers are discussed briefly.

3.6.3.1   GLASS FIBERS

Glass fibers are the most common ceramic fibers to be mass produced today and are divided into long and short fibers. Long glass fibers are produced by either the rod method or the pot method. A long glass fiber is produced by softening one end of a glass rod and drawing it to a fine diameter. Although this is a very simple method, it is not possible to draw very fine fibers (micrometers in size). There are two types of pot method. In one, glass fibers are drawn from multiple nozzles at the bottom of a molten glass tank and are wound together with a binder to form a row of long single fibers. In the other method individual glass fibers are drawn by a jet of high-pressure gas and are collected on a drum, which is subsequently spanned to form rows of long fibers. These two methods are shown schematically in Fig. 3.74.

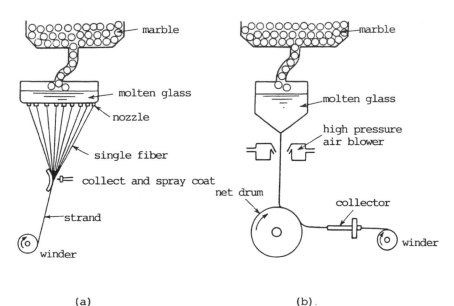

Fig. 3.74 The manufacture of glass fibers by the pot method. (a) Continuous fiber method; (b) staple fiber method

Short glass fibers are also known as glass wool. In the flame blowing method molten glass is blown onto a conveyer by a jet of high-pressure gas and the resulting short fibers are formed into a mat in an oven. In the centrifuge method molten glass is blown out of a container by centrifugal force and the resulting short fibers are formed into a mat in an oven.

3.6.3.2 OPTICAL FIBERS

The most critical issue in the manufacture of optical fibers is the elimination of impurities which are the main causes of light absorption and scattering. Common optical fibers are mostly made of pure glass (called quartz glass) which has been produced from high-purity silicon compounds commonly used for the production of semiconductors. Careful attention has been paid to prevent the introduction of impurities during fiber production.

Optical fibers are commonly made by the CVD method. SiO_2 is formed by the oxidation of $SiCl_4$ with BCl_3 abd $GeCl_4$ which are added to control the refractive indices of optical fibers. As shown in Fig. 3.75, SiO_2 glass is formed on the inside wall of a high-purity quartz tube at elevated temperatures. The quartz tube with SiO_2 deposits is heated to form a preform which is subsequently drawn to form a fine optical fiber. This method is called internal deposition. Thus it is possible to control the distribution of refractive index by

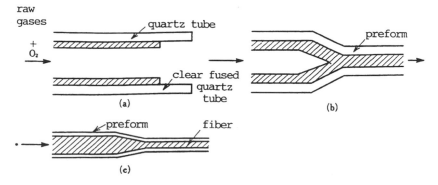

Fig. 3.75   The manufacture of high-purity glass fibers by the CVD method. (a) CVD process; (b) preform formation; (c) fiber drawing

depositing SiO_2 layers with various compositions. In the modified CVD method ultra-fine particles of SiO_2 are deposited in a quartz tube. After deposition the tube is heated by an oxyhydrogen torch to form molten glass along the axial direction of the tube. By this method it is possible to produce optical fibers with more uniform film thickness than by the inside deposition method.

Optical fibers produced by these gas phase methods contain few OH^- groups which are the main cause of an optical loss by absorption. Thus gas phase methods are the main techniques to produce optical fibers. In addition, melt and vapor phase axial deposition methods have been developed. Optical fibers have been used extensively for optical telecommunication and medical applications.

3.6.3.3   CERAMIC FIBERS

Ceramic fibers produced by the sol–gel method have attracted much attention because of their simple manufacturing processes. Gel fibers are produced by spinning a viscous sol with a viscosity above around 10 P (0.1 Ns/m^2) and are heat treated to form either amorphous or crystalline fibers. Not only fibers of simple oxides such as SiO_2, Al_2O_3, ZrO_2, TiO_2, but also those of complex oxides such as $PbTiO_3$ and $YBa_2Cu_3O_{7-x}$ have been produced by the sol–gel method.

A method which uses organosilicon compounds as precursors has also been developed. For example, polycarbosilane which is solid at room temperature is melted at high temperatures and spans to form fibers. When the fibers are treated at elevated temperatures to form bridging bonds at the fiber surfaces and are heat treated at 1200–1500°C, the fibers of β-SiC are formed by thermal decomposition, which is commercially available under the brand name

Nicalon. The fibers of silicon nitride and a Si–Ti–C–O compound have been produced similarly from polysilaxane and titanosiloxane, respectively. The latter fibers are sold commercially under the brand name Tilano.

3.6.4 Porous Bodies

Substances which have pores ranging from a few Ångstroms to a few millimeters are called porous bodies. Although this class of materials includes zeolites which have open crystal structures, porous bodies, in general, imply those produced artificially with a large number of pores. Porous glasses, bubble ceramics and honeycomb structures have been developed for applications as filters, adsorbates, catalyst carriers and thermal insulators (Fig. 3.76).

3.6.4.1 POROUS GLASSES

As discussed in Section 3.2.6, porous glasses are produced by dissolving by acid one of the glass phases which are formed by spinodal decomposition. Porous glasses with various pore sizes have been produced by varying conditions for phase separation. Porous glasses with open pores 10–3000 Å in diameter have been produced and used as enzyme carriers and molecular sieves.

3.6.4.2 POROUS CERAMICS

Porous ceramics have also been produced. Bubble ceramics with 70–90% porosity have been obtained by impregnating ceramic slurries into polymeric sponges such as polyurethane sponges. The composite bodies are heated to decompose and/or vaporize the polymer and sintered to form porous ceramics. Porous ceramics of alumina and aluminosilicates have been developed for applications in thermal insulation, filtration of diesel exhaust fumes and water treatment.

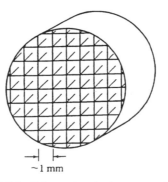

~1 mm

Fig. 3.76 Schematic of a honeycomb structure

Green forms of honeycomb structures are produced by the extrusion method, which are subsequently sintered. It is also possible to form honeycomb structures with a combination of flat plates and corrugated spacers. Honeycomb structures made of cordierite which has a very low thermal expansion coefficient have been used extensively as catalyst carriers for NO_x and hydrocarbon conversion in auto-emission control.

Other methods have also been developed to produce porous ceramics. They include methods which utilize the gelation reaction of polysaccharides such as alginine acid (Fig. 3.77), hollow balloons and biostructures such as wood and coral.

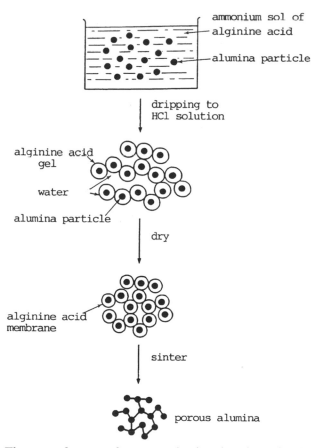

Fig. 3.77 The manufacture of porous alumina by the sol–gel method. (From H. Ichinose, A. Kawahara and H. Katsugi, *New Ceramics*, No. 3, 105, 1992)

3.6.5 Composite Bodies

By forming composite bodies with more than two compounds, it is possible to discover unique characteristics which cannot be obtained from each compound alone. The formation of composite bodies can give (1) a complementary effect which arises due to the complementary strengthening of weaknesses of each compound, (2) an enhancement effect because of new geometries created and (3) a multiplication effect by interfacial interactions. Composite bodies have been developed to satisfy a variety of needs.

3.6.5.1 DISPERSION OF PARTICLES

The dispersion of insoluble particles in metal matrices is known in metallurgy as dispersion strengthening. Similarly, dispersion of SiC, ZrO_2 and Al_2O_3 has been employed to strengthen ceramics. Composite bodies made by the dispersion of nanometer size particles are called nanocomposites, which are expected to show superior fracture toughness.

Dispersion of particles has been used not only to strengthen structural materials but also to derive other functional materials. Examples are rubber magnets and flexible piezoelectric materials made by the dispersion of ceramic particles in rubber, low-temperature IC substrates made by the dispersion of alumina in a glass matrix and particle-dispersed glasses made by the dispersion of CdS and coloring particles in glasses.

3.6.5.2 MULTILAYER CERAMICS

Multilayer ceramics are good examples which have taken advantage of the new geometries created by the formation of composites and include multilayer ceramic capacitors, multilayer chip varistors and multilayer ceramic actuators. These multilayer ceramics are a combination of metallic and ceramic thin layers and are unique in ceramic composites.

The capacitance of a capacitor depends on its geometry. The capacitance made of two parallel plates is given by

$$C = \varepsilon_0 \varepsilon_r (S/d)$$

where C is the capacitance, ε_0 the permittivity of the vacuum, ε_r the relative dielectric constant of the capacitor material, S the surface area of the capacitor and d the thickness of the capacitor. In order to produce high-capacity capacitors, it is necessary to make capacitors with large surface areas and low thicknesses. The multilayer capacitors not only satisfy this requirement but also make it possible to miniaturize. Processes for the manufacture of multilayer ceramic capacitors are shown schematically in Fig. 3.78. Thin layers

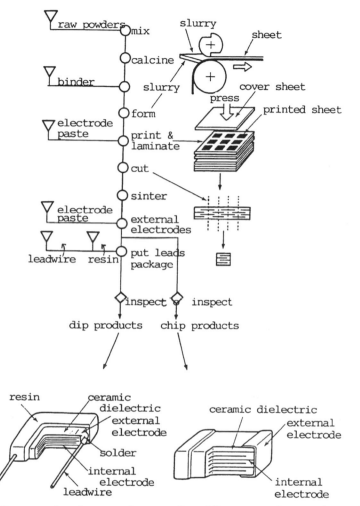

Fig. 3.78 Processes for the manufacture of multilayer ceramic capacitors. (From Y. Yamashita, *Ceramics*, **27**(7), 640, 1992)

of dielectric materials ($BaTiO_3$, etc.) and electrode metal (Ag, Cu, etc.) are stacked alternately to form multilayer capacitors.

3.6.5.3   GRADED STRUCTURES

There is a number of applications which require the use of metal–ceramic, ceramic–ceramic and ceramic–polymer composites. As illustrated by the loss of thermal insulation tiles from the outer skin of a space shuttle due to the large temperature gradient (1000 K) during re-entry, a simple combination of two materials may not satisfy stringent requirements in various applications. Thus attempts have been made continuously to change the composition and structure of an interface.

Examples of graded structures are as follows: SiC/C films on graphite substrates by the CVD method, ZrO_2/Ni–C–Al–Y protective films by the plasma spray method, and ZrO_2/Ni for fuel cells by the slurry method. Graded structures have been sought in both structural and other functional applications (Fig. 3.79).

3.6.5.4 INFILTRATION

Composites are produced by infiltrating other materials in pores and voids of porous bodies made of powders and fibers. Infiltration has been attempted by various methods, which include infiltration of slurries through porous bodies and deposition of precipitates from aqueous solutions and of solids by gaseous reactions (chemical vapor infiltration), which is a derivative of the chemical vapor deposition method.

Since the CVI method relies on the deposition of precipitates in fine pores of porous bodies, it is difficult to form uniform composites simply because it is not possible to continue the deposition of precipitates once gas flow is prohibited by the formation of precipitates. In order to overcome this difficulty, a number of improvements have been attempted and some are shown schematically in Fig. 3.80. The pulsed CVI method in Fig. 3.80(e) has been used to improve the infiltration of reaction gases and the exhaustion of reaction products by periodic evacuation and pressurization of porous bodies which are maintained at a constant temperature. C/C and SiC/SiC composites have been produced from carbon and SiC preforms by the CVI method.

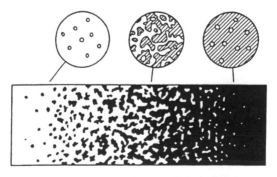

Fig. 3.79 Schematic microstructure of a graded function material. (From Japan Composite Materials Society (ed.), *Dictionary of Ceramic Composite Materials*, Agne Shofu Publishing Inc., 1990, p. 183: reproduced with permission)

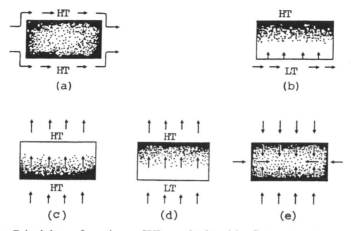

Fig. 3.80 Principles of various CVI methods. (a) Constant temperature; (b) temperature gradient; (c) forced flow CVI; (d) combination of (b) and (c); (e) pulsed gas flow. HT—high temperature; LT—low temperature. (From Japan Composite Materials Society (ed.), *Dictionary of Ceramic Composite Materials*, Agne Shofu Publishing Inc., 1990, p. 62: reproduced with permission)

Table 3.16 Methods to produce composite bodies using long fibers

Method	Brief description
Slurry	Coat fibers with slurry, cut, form and sinter
Sol–gel	Dip fibers in a sol of metal alkoxide, convert to a gel, dry and sinter
CVI	Infiltrate reaction gases through fibers and fill gaps by solid precipitates
Thermal decomposition	Infiltrate a ceramic precursor[a] in fibers and form precipitates in voids by thermal decomposition

[a]Raw materials such as polycarbosilane and polysilaxane, which form ceramics such as SiC and Si_3N_4 by thermal decomposition.

3.6.5.5 FIBER REINFORCEMENT

Fiber reinforcement of materials has been with us for some time. Various products which are reinforced with glass and carbon fibers such as sliding-door papers, skis, tennis rackets, fishing rods and sailing yachts have been produced commercially. In addition, reinforcement of metals, glasses, polymers and ceramics by ceramic fibers such as SiC, alumina and zirconia has been developed.

Long fibers have been used to make reinforced composites and some of the reinforcement methods are listed in Table 3.16. Whiskers which have their dimensions between particles and fibers have also been used to make reinforced composites.

4

PHYSICAL PROPERTIES
OF CERAMICS

Unlike metals and plastics, ceramics are hard and do not combust or rust. These fundamental characteristics are the reasons ceramics are chosen as superior mechanical and structural materials. Progress in fabrication technologies of ceramics has made it possible to turn many substances into materials. Today's ceramics have a wide variety of superior physical characteristics such as electromagnetic, optical and biochemical properties. In this chapter, the prototypic physical properties of ceramics will be discussed.

4.1 THERMAL PROPERTIES

4.1.1 Melting Point

Compared to other competing materials such as polymers and metals, ceramics are highly resistant to heat. In order to exhibit any useful physical functions at elevated temperatures, a material must have this property. Thus it is essential to select ceramics with high melting points for high-temperature applications. Corrosion resistance is another characteristic which is desirable at elevated temperatures. Even if a material is highly resistant to heat, the material cannot be used at elevated temperatures unless it is also corrosion resistant. Compared to metals, most ceramics are stable in oxidizing atmospheres. Some ceramics exhibit excellent electrical properties at elevated temperatures, others excellent mechanical properties at high temperatures and thus they are utilized as high-temperature materials with high hardness and strength.

The melting point of an inorganic material has a strong correlation with its chemical bonding. In general, the melting point increases in the following order: molecular crystals, metallic crystals, ionic crystals (alkali halides and oxides) and covalently bonded crystals (oxides, nitrides, and carbides). The melting point has a stronger correlation with the formation energy of a solid, which is the lattice energy of an ionic crystal.

Melting points of various oxides are listed in Table 4.1. The formation energy of a solid corresponds closely to its crystal structure and exhibits the following trends in oxides:

(1) The higher the coordination number of a cation, the higher the melting point of an oxide.
(2) The closer the ratio between a cation and an anion to unity, or the simpler the chemical composition, the higher is the melting point. This statement corresponds to the fact that when two oxides have cations with an identical coordination number, the oxide whose anions have a higher coordination number has a higher melting point.
(3) The smaller the deviation from the ideal ionic radius ratio, the higher the melting point. This trend has been observed in alkaline earth halides.
(4) The closer the atomic structure of a cation to that of rare gases, or the more stable the atomic valence, the higher the melting point.
(5) The closer the coordination between cations and anions to spherical symmetry, the higher the melting point.

Trends (1) and (2) arise from the fact that oxides which have crystal structures with higher coordination numbers have higher packing densities and thus higher formation energies. Trends (3) to (5) indicate that compounds which have more stable and more homogeneous chemical bonds and crystal structures tend to have higher melting points. Melting points of carbides and nitrides are shown in Fig. 4.1 together with those of oxides.

Table 4.1 Melting points of oxides (°C)

Coordination number	M : X 1 : 1	2 : 3	1 : 2	3 : 4
3		B_2O_3 450		
4	BeO 2410–2573 ZnO sublimation		SiO_2 1723	
6	MgO 2852 CaO 2614 SrO 2420 BaO 1918 NiO 1950	Al_2O_3 2050 Fe_2O_3 1570	TiO_2 1716–1870 SnO_2 1391–1630	$MgAl_2O_4$ 2135 Fe_3O_4 1594 Mn_3O_4 1564
8			ThO_2 3220	
Other		Y_2O_3 2410–2458 Bi_2O_3 825		

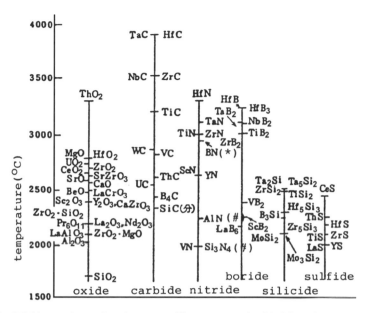

Fig. 4.1 Melting points of various metallic compounds. *Sublimation; # decomposition. (From Agency of Science and Technology (ed.), *Status and Future Prospects of Materials Technologies*, 1974, p. 251)

4.1.2 Thermal Conductivity

In ceramics heat propagates via lattice vibration, namely thermal conduction by phonons. Although electrons can also contribute to thermal conduction, their contribution to thermal conduction in ceramics is quite small. At high temperatures heat can also propagate by the radiation of photons. The conduction of heat can be expressed quantitatively by thermal conductivity, which is a product of the specific heat, carrier mobility (the sonic speed for phonon conduction) and the mean-free path of the carrier. In general, the thermal conductivity of ceramics and organic polymers is smaller than that of metals.

Factors which influence thermal conduction in ceramics are the type of chemical bonding, purity, and density. High thermal conductivity is observed in crystals with (1) strong bonds, (2) a high packing density of atoms in a crystal structure, (3) a high degree of crystallographic symmetry, (4) high crystallinity (glasses have poor crystallinity), and (5) light elements. Crystals with these characteristics have high thermal conductivity because of high mobility and the large mean-free path of phonons.

The propagation of phonons is distributed by lattice defects in crystal structures, pores, grain boundaries, and precipitates in microstructures. As a consequence, thermal conductivity is reduced by the presence of these defects. In addition, it is known that the formation of a solid solution also reduces thermal conductivity.

Substrate materials for integrated circuits (IC) are good examples of materials which must have high thermal conductivity. With increasing integration, more heat is generated. In order to maintain proper operation of ICs, it is necessary to dissipate heat by employing substrate materials with high thermal conductivity. In general, polymeric materials are less thermally conductive and thus cannot be used as substrate materials for highly integrated ICs. Polycrystalline alumina has been used extensively as a substrate material for ICs. In addition, SiC and AlN, which have higher thermal conductivity than alumina, have been used commercially as IC substrates. Recently, diamond, which has the highest known thermal conductivity, has been introduced commercially as an IC substrate material (see Section 4.2.1).

4.1.3 Thermal Expansion

Thermal expansion plays a key role when a combination of two or more materials are used under thermal transients. At the interface between two materials with different thermal coefficients of expansion, materials with higher thermal coefficients of expansion experience a compressive stress during heating and those with lower thermal coefficients of expansion experience tensile stress. As a consequence, microcracks are formed during thermal cycles (heating and cooling), which alter thermal as well as mechanical characteristics and eventually lead to fracture.

As shown in Fig. 4.2, thermal expansion arises due to non-symmetry in an atomic potential which is the sum of repulsive and attractive forces. The lattice vibration increases with increasing temperature (for example, from T_0 to T_1). Simultaneously, the average atomic distance also increases. The non-symmetry in the atomic potential is larger with Coulombic attraction than with covalent bonding. Thus, in general, ionically bonded substances have higher thermal coefficients of expansion than covalently bonded ones. Lattice vibration propagates uniformly throughout the crystals with high packing density. On the other hand, thermal expansion in less dense crystals can be accommodated indirectly by a change in bonding angle. Hence, thermal expansion is higher in denser crystals. Furthermore, crystals with higher coordination numbers and weaker chemical bonding tend to have higher thermal coefficients of expansion. Thus, crystals which have a low melting point tend to have high thermal coefficients of expansion. Thermal coefficients of expansion of various substances are listed in Table 4.2 together with their crystallochemical characteristics. SiO_2 glass, which has a less dense structure with a coordination number of four, has a low thermal coefficient of expansion. The addition of Na_2O and CaO makes silicate glasses much denser and increases their coordination numbers. Thus,

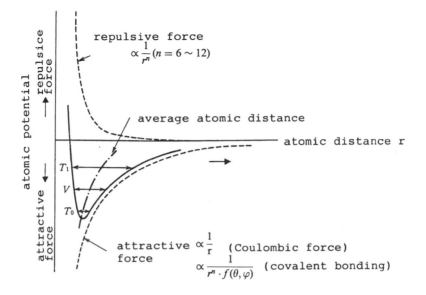

Fig. 4.2 Atomic potential versus atomic distance

Table 4.2 Thermal coefficients of expansion and crystal structures of various substances

Substance	Coordination number	Product of elect. charges	Bond[a] strength	Difference in electro-negativity	Thermal coeff. of expansion (0–1000°C) (10^{-6} K^{-1})
Diamond	4	16	4	0	~0
SiO$_2$ (glass)	4	8	2	1.7	0.5
SiC	4	16	4	0.7	4.7
BeO	4	$2 \times 2 = 4$	1	2.0	9.0
SiO$_2$–Na$_2$O– CaO (glass)	4 (6–8)	8, 2, 4	2, 0.5, 1	1.7 (2.6, 2.5)	9.0
Al$_2$O$_3$	6	6	1	2.0	8.8
MgO	6	4	0.67	2.3	13.5
NaCl	6	1	0.167	2.1	40.0
ThO$_2$	8	8	1	2.2	9.2
ZrO$_2$ (CaO stabilized)	8	8, 4	1, 0.5	2.1 (2.5)	10.0

[a]Values obtained by dividing the products of electronic charges for the given pairs of atomic bonds by the coordination numbers.

their thermal coefficients of expansion become larger. Silicate minerals which have a large amount of voids in the form of rings and three-dimensional networks such as cordierite ($Mg_2Al_4Si_5O_{18}$) have very low thermal coefficients of expansion (negative thermal expansion along the c-axis). These minerals contain both strong and weak bonds and the thermal expansion of stronger bonds can be accommodated by the rotation of ring structures.

4.2 ELECTRICAL PROPERTIES

Metals are good electrical conductors and also exhibit important magnetic properties. Polymers are, in general, electrical insulators. However, ceramics exhibit a wide variety of important electrical properties which include electrical insulation, magnetic and dielectric properties, and electrical conductivity (semiconductivity). These various electrical properties stem from a wide variation in chemical composition, crystal structure, microstructure and synthesis method. This variety is unique to ceramics.

The capability of controlling electrical properties by the formation of a solid solution is one of the unique features of ceramics. In principle, electrical properties are dictated by atoms and their combinations, namely crystal structure. Thus electrical properties of a pure solid are determined by these two factors. However, it is possible to alter electrical properties drastically by substituting one kind of atom with another, namely by forming a solid solution. By controlling the degree of substitution it is also possible to control the amount of changes in electrical properties. Examples of this unique feature are the changes in electrical conductivity due to the formation of lattice defects by the addition of minor additives and the changes in magnetic characteristics of spinel ferrites by the controlled distribution of magnetic atoms.

The possibility of controlling electrical properties by manipulating microstructures is another unique feature of ceramics. Polycrystalline sintered bodies are made of complicated microstructures which include grains, grain boundaries, pores and surfaces and thus exhibit significantly different electrical properties from those of single crystals. A number of unique ceramic materials which have been developed by manipulating microstructures include PTC thermistors, ZnO varistors and boundary layer capacitors (by manipulating grain boundary characteristics) and gas sensors and various thin-film semiconductors (by manipulating surface characteristics.

4.2.1 Electrical Insulation

Ceramics whose electrical conductivity is less than $10^{-8} \, \Omega^{-1} \, cm^{-1}$ are called electrically insulating. The band gaps of these ceramics are quite large and are in the range 2–8 eV (see Section 4.2.2).

Typical ceramic insulators are common porcelain, which is made of clay, feldspar and quartz, alumina (Al_2O_3)–mullite ($3Al_2O_3 \cdot 2SiO_2$), steatite ($MgO \cdot SiO_2$)–forsterite ($2MgO \cdot SiO_2$), beryllia (BeO), nitrides such as Si_3N_4 and various silicate glasses. Electrical characteristics of these insulators are listed in Table 4.3.

In order to discuss the electrical insulation of ceramics adequately, it is necessary to consider the contribution of a capacitive component and ionic and surface conductions to electrical resistance. For materials with high capacitance, C, it is not possible to maintain electrical insulation at a high frequency because of the decrease in capacitive resistance R_c ($R_c = 1/j\omega C$, where ω is the angular frequency). Thus, it is necessary to employ ceramic materials with a low dielectric constant for microelectronic applications such as IC substrates and packages which require good electrical insulation at high frequencies.

The electronic conductivity of insulating ceramics is negligible near room temperature and their electrical conduction arises mainly due to the diffusion of ions. In general, ions with a small valence and a small ionic radius diffuse much faster than those with a large valence and ionic radius. Hence, alkali ions, which are present as impurities, tend to contribute significantly to the overall electrical conductivity of a ceramic material. Therefore, the concentration of alkali ions is kept as low as possible in ceramic insulating materials. Surface conductivity arises mainly due to water molecules which are present as a result of absorption and condensation. The presence of pores at the surface and impurities which dissociate in water to form mobile ions tends to increase surface conductivity. Planarization and glazing of ceramic surfaces have been employed to minimize this.

In addition to superior electrical characteristics (high resistivity, high dielectric breakdown voltage and low dielectric loss), ceramic insulators must have superior thermal, mechanical and chemical properties. Thus, a variety of ceramic insulators are employed in various applications. For example, insulators for high-voltage transmission lines must have high dielectric breakdown voltage and high mechanical strength and are made mainly of mullite and alumina. Spark plug insulators for internal combustion engines are exposed to high temperature and high pressure and thus must have good thermal shock resistance. Alumina is the preferred choice for application to spark plugs. IC substrates must have high thermal conductivity, high electrical insulation at high frequencies and high dimensional tolerance and are made of high-density sintered alumina or single-crystal alumina (sapphire).

Monolithic SiC is a semiconductor with an electrical resistivity of 10^2 to $10^5\,\Omega\,cm$. However, SiC hot-pressed with around 2 wt% of BeO exhibits high electrical resistivity with excellent thermal conductivity which is about a factor of ten higher than alumina. Although grains in this hot-pressed SiC have low electrical resistivity, the grain boundaries have high electrical resistivity due to

Table 4.3 Physical properties of ceramic insulators

Insulator	Alumina	Alumina	Forsterite	Mullite	Beryllia	Aluminum nitride	Silicon carbide
Major component	96% Al_2O_3	99.5% Al_2O_3	$2MgO \cdot SiO_2$	$3Al_2O_3 \cdot 2SiO_2$	99% BeO	AlN	SiC (+BeO)
Specific gravity (g cm^{-3})	3.75	3.90	2.8	3.1	2.9	3.3	3.2
Bending strength (MPa)	340	490	140	180	190	400–500	450
Young's modulus (GPa)	300	380	–	100	320	280	400
Thermal properties TCE[a] (10^{-6} K^{-1})	6.7	6.8	10	4.0	6.8	4.5	3.7
Thermal conductivity (25°C) (W/mK)	22	31	3	4	240	100–260	270
Electrical properties							
Breakdown voltage (20°C) (kV/mm)	14	15	13	13	15	15	0.1–0.3
Volume resistivity							
20°C (Ωcm)	$>10^{14}$	$>10^{14}$	$>10^{14}$	$>10^{14}$	$>10^{14}$	10^{13}–10^{15}	$>10^{13}$
500°C (Ωcm)	4.0×10^9	3×10^{12}	1×10^{10}	–	1×10^{13}	–	–
Dielectric constant (1 MHz, 20°C)	9.0	9.8	6.0	6.5	6.8	8.8	42
Major applications	Spark plugs, IC substrates	Thin film IC substrates	Common electrical components	Excellent thermal shock resistance	High thermal conductivity substrates	High thermal conductivity substrates	High thermal conductivity substrates

[a]Average thermal coefficient of expansion (TCE) between 25°C and 300°C (TCE for SiC between 25 and 400°C)
(From *Applications of Emerging Materials*, pp. 67 and 69, Society of Industrial Survey (1990) and *Ceramic Data Book, 1985*, p. 369, Society of Industrial Product Technologies (1985).)

the depletion of carriers. BeO dissolves in SiC with difficulty. However, the addition of BeO promotes the sintering of SiC and thus minimizes the scattering of phonons.

AlN was known theoretically to have high thermal conductivity. However, it was not possible to make AlN with high thermal conductivity because of its poor sinterability and the presence of oxygen and silicon as impurities. Today high thermal conductivity AlN can be sintered in reducing atmospheres at high temperatures with high-purity AlN starting powders and a few weight per cent of Y_2O_3 or CaO as sintering aids. Not only does the additive promote the sintering of AlN, but the resulting $YAlO_3$ and $Y_3Al_5O_{12}$ also stimulate the removal of oxygen from AlN grains, which leads to high thermal conductivity. As shown in Fig. 4.3, the thermal conductivity of AlN increases with decreasing amounts of oxygen in raw AlN powders.

4.2.2 Semiconductivity

4.2.2.1 BAND STRUCTURE

The concept of band structure is very effective in understanding electrical conduction in solids. There are two ways to interpret the formation of band structures in solids:

(1) Explaining band structure by invoking the interaction of electron energy levels of individual atoms. When atoms are brought closer together, they interact with each other. As a consequence, their electron energy levels are

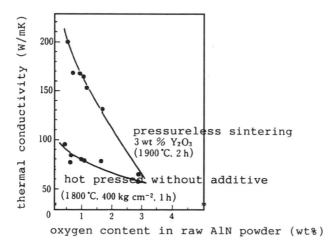

Fig. 4.3 Thermal conductivity of polycrystalline AlN as a function of oxygen content in raw AlN powder. (From F. Miyashiro *et al.*, ©1990 *IEEE Trans. CHMT*, **13**, 313)

distorted and form multiple energy sublevels with very small difference in energy. With increasing interaction these energy sublevels form semi-continuous energy bands as shown in Fig. 4.4.

(2) Explaining band structure by invoking the motion of electrons in a periodic potential formed by a crystal lattice. Energies of free electrons are proportional to k^2, where k is the wave number of a wave equation. When electron potentials oscillate with a period of lattice constant a, discontinuities in the $E–k$ curve occur at $k = n\pi/a$ ($n = \pm1, \pm2, \ldots$). Electron waves are reflected at these k and thus form bands whose energies electrons are allowed to possess. Electrons are filled first at the lowest energy level and subsequently at progressively higher energy levels. The degree of electron filling is expressed by the Fermi level E_F. At absolute zero temperature energy levels up to E_F are filled with electrons, but energy levels above E_F are completely empty. At temperatures above absolute zero, some electrons are excited from energy levels lower than E_F by thermal excitation and occupy energy levels higher than E_F. The probability $f(E)$ of finding electrons with energy E at temperature T can be given by

$$f(E) = 1/[1 + \exp{(E - E_F)/kT}] \tag{4.1}$$

This is called the Fermi–Dirac distribution. Thus E_F can be defined by the energy level with the probability of $1/2$. In general, the highest energy band filled with electrons is called the valence band and the next energy band which will be filled with electrons for any reason is the conduction band. The energy levels between these bands are called forbidden bands and their magnitude the band gaps (Fig. 4.5).

It is possible to classify materials based on the band structure. Band structures of a metal, a semiconductor and an insulator are shown in Fig. 4.6.

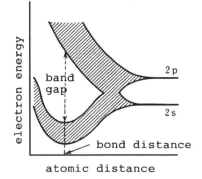

Fig. 4.4 Energy band based on the atomic interaction model

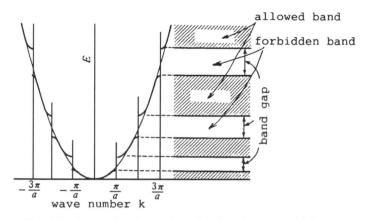

Fig. 4.5 Energy band based on the free electron model

Insulators have large band gaps. At room temperature the valence band is filled with electrons and no electrons are present in the conduction band. Thus insulators do not have any carriers for electrical conduction. However, some electrons are excited from the valence to the conduction bands at elevated temperatures by thermal motion and thus insulators exhibit some electrical conductivity. The highest energy band of metals are partially filled with electrons which can move very easily in the band. Semiconductors are situated between metals and insulators and have smaller band gaps than insulators. Semiconductors with few impurities and equal numbers of electrons and electron holes created by thermal excitation are called intrinsic. On the other hand, those semiconductors with electrons introduced by non-stoichiometry or

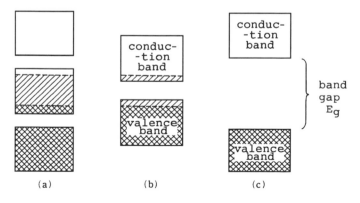

Fig. 4.6 Classification of materials based on the energy band. (a) Metal; (b) semiconductor; (c) insulator

impurities in the conduction band are called *n*-type semiconductors and those with electron holes in the valence band are called *p*-type.

Ionic crystals of representative elements have large band gaps and are therefore insulators. Ions of representative elements have stable electron orbitals which are very close to those of rare gases. Since the excitation of electrons from the valence to the conduction band corresponds to the removal of electrons from anions to cations, the excitation of electrons makes the electron configurations of ions different from those of rare gases. Such an excitation is also quite disadvantageous energetically. Thus MgO and Al_2O_3 have large band gaps of 8–10 eV.

Covalently bonded crystals with a coordination number of four have valence bands made of sp^3 mixed orbitals and conduction bands of sp^3 anti-bonding orbitals. The main difference between these two orbitals is the direction of those four orbitals. As a consequence, the differences in energy between them are quite small. The highest band gap in this class of crystals is that of diamond (about 5 eV).

Band gaps of substances made of representative elements are shown in Fig. 4.7 and those of covalently bonded crystals tend to decrease with increasing atomic number: for example, diamond > SiC > Si > Ge. Band gaps of compounds which have an identical number of total electrons and belong to same period of the periodic table increase as follows:

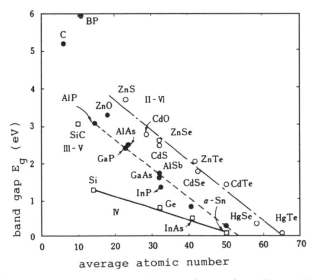

Fig. 4.7 Correlation of band gap with atomic number. (From K. Fueki *et al.*, *Chemistry of Electrical Materials*, Maruzen, 1981, p. 29)

$$Ge < GaAs < ZnSe < CuBr$$

The above order reflects the trend that the difference in electronegativity of representative elements increases as indicated and thus ionicity of bonding increases accordingly. Oxides such as ZnO which are partially bonded covalently tend to have smaller band gaps.

4.2.2.2   ELECTRICAL CONDUCTIVITY AND THE FERMI LEVEL

The electrical conductivity, σ, of a semiconductor can be given by

$$\sigma = en_e\mu_e + en_h\mu_h \tag{4.2}$$

where n_e, n_h, μ_e, μ_h, and e are the numbers of electrons and electron holes in unit volume, the mobilities of electrons and electron holes and electronic charge, respectively. In intrinsic semiconductors $n_e = n_h$. In n-type semiconductors $n_e \gg n_h$ and in p-type semiconducotrs $n_e \ll n_h$.

The Fermi level of an intrinsic semiconductor can be given by

$$\begin{aligned} E_F &= (E_c + E_v)/2 + (kT/2)\ln(N_v/N_c) \\ &= (E_c + E_v)/2 + (3kT/4)\ln(m_h^*/m_e^*) \end{aligned} \tag{4.3}$$

where E_c and E_v are the energies at the bottom of the conduction band and at the top of the valence band, respectively. N_v and N_c are called the effective state densities of the conduction and valence bands, respectively, and can be given by

$$N_c = (2\pi m_e^* kT/h^2)^{3/2}$$
$$N_v = (2\pi m_h^* kT/h^2)^{3/2}$$

where m_e^* and m_h^* are the effective masses of an electron and electron hole, respectively. When $m_e^* = m_h^*$, E_F is at the middle of the forbidden band. In many semiconductors, $m_e^* < m_h^*$. Thus E_F increases with increasing temperature.

The concentrations of electrons and electron holes, n_e and n_h, can be given by

$$n_e = n_h = AT^{3/2} \exp\{-(E_c - E_v)/2kT\} \tag{4.4}$$

where $A = 2(2\pi mk/h^2)^{3/2}$ with the assumption that $m = m_e^* = m_h^*$. Since electron mobility has a dependence of $T^{-3/2}$, which cancels out with $T^{3/2}$ in Eq. (4.4), the electrical conductivity, σ, can be given by

$$\sigma = \sigma_0 \exp\{-(E_c - E_v)/2kT\} \tag{4.5}$$

where σ_0 is the constant. Thus a logarithmic plot of σ versus $1/T$ gives a straight line from whose slope it is possible to determine the band gap E_g $(= E_c - E_v)$.

Non-stoichiometric defects and impurities in a crystal form a variety of levels in the forbidden band. Electrons are excited from and electron holes are created by the levels. (See Section 2.6 for details.) Defects and impurities which excite electrons are called donors. In n-type semiconductors, electrons from donors are the majority charge carriers. As indicated in Fig. 4.8, donor levels are located below the bottom edge of a conduction band. On the other hand, defects and impurities which accept electrons and thus create electron holes in the valence band are called acceptors. In p-type semiconductors, electron holes from acceptors are the majority charge carriers. Acceptor levels are located above the top edge of a valence band. As a consequence, the formation energies of electrons in n-type semiconductors and electron holes in p-type semiconductors are smaller than those of intrinsic semiconductors. The Fermi level of an n-type semiconductor can be expressed by substituting E_v and N_v in Eq. (4.3) with E_d and N_d and is given by

$$E_F = (E_c + E_d)/2 + (KT/2)\ln(N_d/N_c) \tag{4.6}$$

Similarly, the electrical conductivity can be obtained by the substitution of Eq. (4.5) and is given by

$$\sigma = \sigma_0 \exp\{-(E_c - E_d)/2kT\} \tag{4.7}$$

On the other hand, the Fermi level and electrical conductivity of a p-type semiconductor can be obtained by substituting E_c and N_c in Eqs (4.3) and (4.5) with E_a and N_a and are given by

$$E_F = (E_a + E_v)/2 + (kT/2)\ln(N_v/N_a) \tag{4.8}$$

$$\sigma = \sigma_0 \exp\{-(E_a - E_v)/2kT\} \tag{4.9}$$

Even in n- and p-type semiconductors electrons are excited from the valence band to the conduction band at high temperatures. In this situation the Fermi level and electrical conductivity can be expressed identically to those of intrinsic semiconductors. Thus, as indicated in Fig. 4.9, the temperature dependence of electrical conductivity exhibits two distinct regions. The low-temperature region is called extrinsic. Since the majority of carriers are electrons and electron holes created by defects and impurities, the electrical conductivity depends strongly on the concentration of defects and impurities. The high-temperature region is called intrinsic. The electrical conductivity is characteristic of a given semiconductor and has a unique slope of $E_g/2k$.

4.2.2.3 $p–n$ JUNCTIONS

Interfaces between p- and n-type semiconductors exhibit a rectification phenomenon in which electric current can pass along one direction but cannot pass along the opposite one. Junction transistors and solar cells have been made based on this phenomenon. The band structures of $p–n$ junctions are shown in Fig. 4.10.

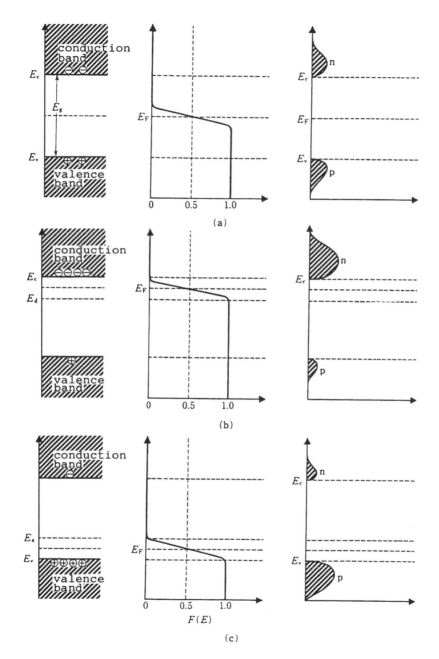

Fig. 4.8 Band structure, Fermi level and electron and electron hole concentrations of various semiconductors. (a) Intrinsic semiconductor; (b) *n*-type semiconductor; (c) *p*-type semiconductor

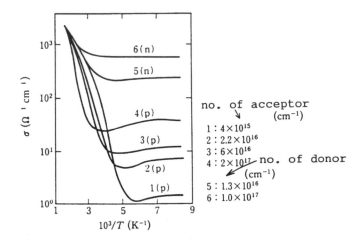

Fig. 4.9 Temperature dependence of electrical conductivity of *n*- and *p*-type semiconductors. (From O. Madelung and H. Weiss, *Z. Naturforsch.*, **9a**, 527, 1954)

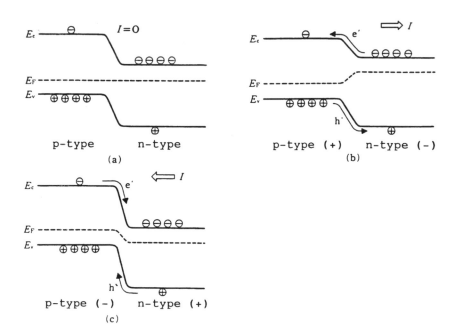

Fig 4.10 Band structure of a *p–n* junctions. (a) no bias; (b) forward bias; (c) reverse bias

When an n-type semiconductor comes into contact with a p-type semiconductor, electrons in the n-type semiconductor migrate to the p-type semiconductor and electron holes in the p-type semiconductor migrate to the n-type semiconductor. As a consequence, the Fermi levels of both semiconductors coincide with each other. This condition is called thermal equilibrium. When a positive bias is applied to the p-type semiconductor (or a negative bias to the n-type semiconductor), the Fermi level of the p-type semiconductor becomes lower. As a consequence, electrons migrate from the n-type to the p-type semiconductor and electron holes migrate from the p-type to the n-type semiconductor. This is called forward bias. Current I increases exponentially with increasing bias V and is given by

$$I = I_0\{\exp(eV/kT) - 1\} \tag{4.10}$$

where I_0 is a constant. When a negative bias is applied to the p-type semiconductor (or a positive bias to the n-type semiconductor), little electric current flows because there are very few electrons in the p-type semiconductor and very few electron holes in the n-type semiconductor. Although Eq. (4.10) is still valid in this situation, electric current never exceeds $-I_0$ because V is negative. This situation is called reverse bias. When reverse bias exceeds a certain high value, electric current starts flowing catastrophically. This is called the dielectric breakdown voltage. The I–V characteristic of a p–n junction is shown in Fig. 4.11.

When a semiconductor is irradiated with photons with an energy larger than its band gap, electrons are excited to the conduction band and thus the electrical conductivity of the semiconductor increases. This phenomenon is called photoconductivity. When a p–n junction is formed with a semiconductor which exhibits photoconductivity, the concentrations of the minority carriers

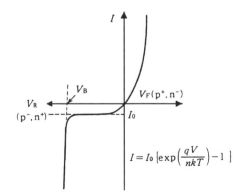

Fig. 4.11 Voltage–current characteristics of a p–n junction. V_F— forward bias voltage; V_R —reverse bias voltage; V_B—dielectric breakdown voltage

increase with the irradiation of light without bias and electric current starts flowing. This current is the source of a solar cell which is made mainly of polycrystalline and amorphous Si. Solar cells are also formed by the electrodeposition of p-type Cu_2S on the surface and grain boundaries of n-type CdS (Fig. 4.12).

4.2.2.4   THERMISTORS

Thermistors, or thermal sensitive resistors, are a class of semiconductors which exhibit a steep change in electrical resistivity with changing temperature. They have been used extensively for temperature measurements and controls.

As indicated in Fig. 4.13, practical thermistors have their electrical resistivity between 10^0 and 10^6 Ωcm. They can be classified into three major categories: NTC (negative temperature coefficient) thermistors whose electrical resistivity decreases with increasing temperature; PTC (postive temperature coefficient) thermistors whose electrical resistivity increases with increasing temperature; and CTR (critical temperature resistor) thermistors which exhibit a drastic change in electrical resistivity at a critical temperature.

NTC Thermistors

NTC thermistors consist mainly of transition metal oxides such as CoO, NiO and MnO and have crystal structures which are very close to that of spinel. These semiconductors exhibit stable electrical characteristics which are immune to the influence of atmosphere such as oxygen on electrical conductivity and are comparatively less affected by the presence of impurities.

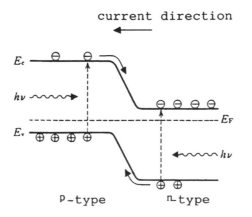

Fig. 4.12   Band structure of a p–n junction

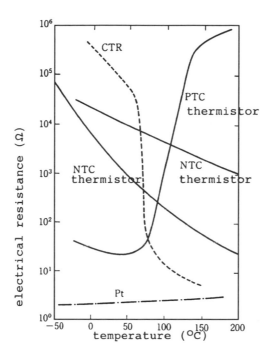

Fig. 4.13 Temperature dependence of the electrical resistance of various thermistors.
H. Futaki 'Thermistor', p. 45, Nikkan-Kogyo-Shimbun (1980)

The temperature dependence of electrical resistance in NTC thermistors can be
given by

$$R = R_0 \exp\{B(1/T - 1/T_0)\} \tag{4.11}$$

where R and R_0 are electrical resistances at temperatures T and T_0, respectively,
and B is the thermistor constant. The relationship between the change in
electrical resistance α for every $1\,°C$ change in temperature and thermistor
constant can be given by

$$\alpha = -B/T^2$$

Commercial thermistors have a B value of between 2000 and 5000 K. A
thermistor with a B of 3600 K has $\alpha = -0.04$ at room temperature and thus
exhibits a change of 4% in electrical resistance per degree Centigrade. High-
temperature thermistors which are utilized for the temperature measurement of
exhaust gas of automobiles at temperatures above $500\,°C$ must have large
temperature coefficients of electrical resistance and must also be stable
thermally without a change in their crystal structure. This class of NTC
thermistors is composed of semiconductors with fluorite structures such as

those of ZrO_2–Y_2O_3 and ZrO_2–CeO_2 systems and with spinel structures such as those of $CoAl_2O_4$ and $Mg(Al, Cr, Fe)_2O_4$.

PTC Thermistors

$BaTiO_3$ has a perovskite structure and exhibits ferroelectricity with a high resistivity of 10^{12} Ω cm around room temperature. When Ba or Ti ions are substituted with cations with an ionic radius very close to those of Ba or Ti ions and with higher atomic valence, $BaTiO_3$ becomes an *n*-type semiconductor. Rare earth elements such as La^{3+} substitute Ba^{2+} ions and cations such as Sb^{5+}, Nb^{5+} and Ta^{5+} substitute Ti^{4+} ions. In order to maintain electrical neutrality, electrons are excited into the conduction band. As a consequence, $BaTiO_3$ becomes an *n*-type semiconductor.

When a polycrystalline body of *n*-type $BaTiO_3$ is heated, it exhibits a sharp increase in electrical resistivity around 120 °C which is close to the Curie point of BaTiO (Figs 4.13 and 2.27). This phenomenon has been observed in polycrystalline $BaTiO_3$, but not in single-crystal $BaTiO_3$, as well as in polycrystalline $BaTiO_3$ specimens which are sintered in reducing atmospheres and are oxidized subsequently to form oxidized grain boundaries.

Along the grain boundaries of $BaTiO_3$ the capture of electrons by acceptors such as adsorbed oxygen ions forms potential barriers whose height can be given by

$$e\phi = e^2 N_s^2 / 2\varepsilon_r \varepsilon_0 N_d$$

where N_s and N_d are the concentrations of acceptors along grain boundaries and donors in a grain matrix and ε_r and ε_0 are the dielectric constant and the permittivity of vacuum, respectively. Since the dielectric constant decreases according to the Curie–Weiss law at temperatures above the Curie point, the height of the potential barrier increases drastically and thus electrical resistivity also rises. It is thought that the increase in the heights of the potential barriers is the main cause of PTC characteristics.

The Curie point can be shifted by forming solid solutions and the shift of the Curie point of $BaTiO_3$ has been shown in Fig. 2.26. Since the resistivity of PTC thermistors increases with increasing temperature, heaters made of PTC thermistors are able to control their power consumption. This function has been utilized extensively in the manufacture of self-controlled heaters.

CTR

As indicated in Fig. 4.13, thermistors made of the V_2O_5–V_2O_4 system exhibit a sharp decrease in electrical resistivity around 70 °C. This phenomenon is due to the semiconductor-to-metal transition which is induced by a minor change in crystal structure.

4.2.2.5 GAS SENSORS

Thin-film or porous semiconductors exhibit drastic changes in electrical conductivity due to absorption of and reaction with gases in the environment, and thus it is possible to measure the composition and concentration of gases by measuring the change in electrical conductivity. This type of gas sensor has been used extensively to detect gas leaks and to control the concentrations of various industrial gases.

The electrical conductivity of semiconductors is affected by the chemisorption of gases, which accompany the transfer of electrons during absorption. For example, the absorption of gas molecules such as oxygen, which forms anions at the surface of an n-type semiconductor, causes the transfer of electrons from the semiconductor to the gas molecules. The electron transfer ionizes the semiconductor and thus creates surface levels in the forbidden band. In order to maintain electrical neutrality, the band is bent and thus a potential barrier is created. As a consequence, the absorption of oxygen on the surface of an n-type semiconductor in air creates a high-resistivity layer which is electron-deficient and a few tens of namometers thick.

When combustible gases such as CO, H_2 and propane come into contact with the surface of semiconductor gas sensors, the amount of oxygen molecules on the surface decreases by the reaction of oxygen with the combustible gases and thus the height of the potential barrier also decreases. The high resistivity of grains, grain boundaries and neck areas of a porous semiconductor is shown schematically in Fig. 4.14. The reduction of potential barrier height accompanies the lowering of barrier heights at grain boundaries as well as the broadening of conductive channel width at the neck areas. As a consequence, overall resistivity decreases drastically. Furthermore, in semiconductors with very fine grains the electrical conductivity of the grain matrix itself is altered. Thus, the sensor characteristics are strongly influenced by the ceramic microstructure.

Ceramic sensors are made of n-type semiconductors such as SnO_2 and ZnO. As shown in Fig. 4.15, the sensitivity of gas sensors is high for highly combustible gases such as alcohol and those gases with a large number of carbon atoms. The selectivity of the gas sensors for specific gases improves with the addition of a catalyst of a noble metal, such as Pt or Pd, and oxide, such as ThO_2, CuO, WO_3, V_2O_5 or MoO_3. It is believed that the catalyst activates gases and enhances the rate of absorption and oxidation of a specific gas. However, the function of catalysts has not been understood fully. In order to improve selectivity, attempts have also been made to employ metal/semiconductor interfaces as well as interfaces between various semiconductors.

The electrical conductivity of porous semiconductors changes significantly through the absorption of moisture at room temperature. This phenomenon has been applied for detecting moisture. When water molecules are physisorbed on the surface, water condenses at concave surfaces and in capillaries. H^+ ions (as well as H_3O^+ ions) formed by the dissociation of water molecules migrate quite freely through the absorbed water and thus electrical

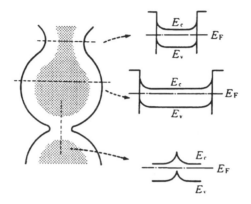

Fig. 4.14 Schematic of porous sintered bodies and their band structures. The surface layer is electrically insulating due to electron deficiency. E_c—energy level at the bottom of the conduction band; E_v—energy level at the top of the valence band; E_F—Fermi level (Reproduced by permission of the Chemical Society, London)

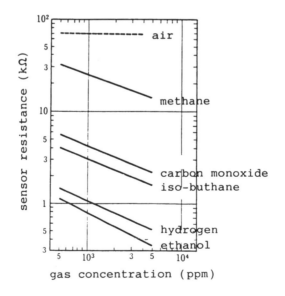

Fig. 4.15 Dependence of prototypic SnO_2 gas sensors on the concentrations of various gases

resistivity decreases. This class of moisture sensors comprises $MgCr_2O_4$–TiO_2, ZnO–Li_2O–V_2O_5 and Fe_3O_4 systems.

4.2.3 Dielectric Properties

4.2.3.1 POLARIZATION

When an electric field is applied to a material, charge carriers such as ions, electrons, and electron holes are displaced and their displacement can be classified into the following two categories: (1) charge carriers migrate macroscopically under the influence of an electric field and thus electric current flows through the material; (2) charge carriers are bound to some specific positions and migrate only microscopically under the influence of an external field. The deviation of centers of gravity of positive and negative charges due to the minor movement of charge carriers is called polarization.

The origins of polarization can be classified into the following;

(1) *Electronic polarization*: polarization caused by the shift of electrons in an atom with respect to nucelei.
(2) *Ionic polarization*: polarization caused by the shift of centers of gravity of positive and negative ions in an ionic crystal.
(3) *Orientation polarization*: molecules such as H_2O and $CHCl_3$ which have permanent dipole moments have a random orientation and the statistical sum of the permanent dipole moments is zero. When an electric field is applied to the material, the permanent dipole moments orient along the direction which give the most stable configuration. The polarization caused by the orientation of dipole moments is called orientation polarization.
(4) *Space charge (interfacial) polarization*: polarization caused by the formation of space charges due to the accumulation of charge carriers in some specific spaces or interfaces in heterogeneous dielectric bodies such as polycrystalline ones.

The magnitude of polarization P can be defined by the dipole moment in unit volume ($C\,m^{-2}$), which is also the area density of polarization charges induced on the surface of a dielectric body. A dipole moment μ ($C\,m$) induced by the application of an electric field E_i to a molecule can be given by

$$\mu = \alpha E_i \tag{4.12}$$

where α is the constant called the polarization coefficient. Assuming that the number of molecules per unit volume is N, the polarization P can be given by

$$P = N\mu = N\alpha E_i \tag{4.13}$$

Since individual molecules are polarized, E_i is different from the external field.

The electric flux density D ($C\,m^{-2}$) of a dielectric body in an external field E is given by

$$D = \varepsilon E = \varepsilon_s \varepsilon_0 E \tag{4.14}$$

where ε, ε_s and ε_0 are the dielectric permittivity, the dielectric constant and the permittivity of vacuum, respectively. D is also the sum of the electric flux density in vacuum and the electric flux density due to the polarization and thus is given by

$$D = \varepsilon_0 E + P \tag{4.15}$$

Therefore, the polarization P is given by

$$P = (\varepsilon_s - 1)\varepsilon_0 E \tag{4.16}$$

P can also be given by

$$P = \varepsilon_0 \chi E$$

where χ is electric susceptibility.

When an electric field is applied to organic molecules or a crystal with a permanent dipole moment of μ_0, polarization P becomes a function of absolute temperature due to the balance between the electric field and thermal motion. When $\mu_0 E \ll kT$, the polarization is given by

$$P = N\mu_0^2 E / 3kT \tag{4.17}$$

where k is the Boltzmann constant. Thus the polarizability of a single molecule, α_0, can be given by

$$\alpha_0 = \mu_0^2 / 3kT \tag{4.18}$$

The four types of polarization discussed above do not take place instantaneously upon the application of an external field. Instead, each polarization occurs with a characteristic delay. This phenomenon is called dielectric relaxation. The time required to reach thermal equilibrium is called the relaxation time and its inverse is called the relaxation frequency. The relaxation time decreases in the following order: space charge polarization, orientation polarization, ionic polarization and electronic polarization. The polarizability and dielectric loss are shown schematically as a function of frequency in Fig. 4.16.

The dielectric permittivity under an alternating electric field can be given by

$$\varepsilon = \varepsilon' - j\varepsilon'' \tag{4.19}$$

where ε' and ε'' are the real and imaginary parts of the dielectric permittivity, respectively. When an alternating electric field E of a frequency f is applied, the

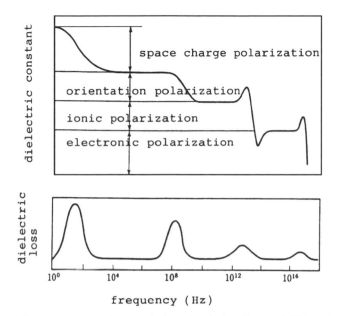

Fig. 4.16 Frequency characteristics of polarizability and dielectric loss

phase angle of the electric flux density lags behind that of the electric field due to the finite speed of polarization. The delay angle δ is given by

$$\tan \delta = \varepsilon'' / \varepsilon' \tag{4.20}$$

Since the electric power loss per unit time (which is called the dielectric loss) is proportional to $2\pi f \varepsilon_0 \varepsilon' E^2 \tan \delta$, $\tan \delta$ or ε'' is designated as the characteristic value for the dielectric loss. As indicated in Fig. 4.16, the dielectric loss exhibits a maximum when the frequency of the external electric field coincides with the relaxation frequency of a given polarization.

4.2.3.2 CLASSIFICATION OF DIELECTRICS

The dielectric properties of a material are intimately related to atoms which comprise the material and its crystal structure. All dielectric materials can be classified into 32 symmetry groups. Dielectric materials which belong to 11 symmetry groups with a center of symmetry and to a group designated by 0 are classified as paraelectric. Dielectric materials which belong to the remaining 20 symmetry groups without a center of symmetry are called piezoelectric. When these dielectric materials are either stressed or strained, the center of gravity of positive and negative charges deviates and thus the polarization occurs. Ten groups of dielectric crystals of the 20 symmetry groups which exhibit piezoelectricity have especially low symmetry and exhibit polarization without the application of either electric field or stress, which is called spontaneous polarization.

In general, the absorption of molecules and ions cancels out spontaneous polarization, but a change in temperature results in polarization to crystal surfaces due to thermal expansion and the change in thermal vibration. This phenomenon is called pyroelectricity. The direction of spontaneous polarization of some pyroelectric crystals can be altered by the application of external electric fields. This phenomenon is called ferroelectricity. Thus, as indicated in Fig. 4.17, ferroelectric crystals exhibit both pyroelectricity and piezoelectricity (see Table 2.9).

Barium titanate and its solid solutions, namely (Ba, Sr, Pb, Ca) (Ti, Zr, Sn)O_3, comprise the majority of ferroelectric ceramics. On the other hand, lead zirconate–titanates Pb(Ti, Zr)O_3 with large piezoelectric constants comprise the majority of piezoelectric ceramics. Ordinary dielectric ceramics which do not exhibit either ferroelectricity or piezoelectricity are quite stable with respect to changes in electric field, mechanical stress, and temperature and are thus used widely as electric insulators and capacitor materials, which require stable dielectric characteristics.

As indicated in Fig. 4.18, there are several types of polarization, P induced by an external field E. Paraelectric materials (a) do not exhibit any hystereses. On the other hand, ferroelectric (b), antiferroelectric (c) and ferrielectric (d) materials exhibit a variety of hysteresis curves. Magnetic materials do not exhibit the types of hysteresis indicated by Fig. 4.18(c) or (d).

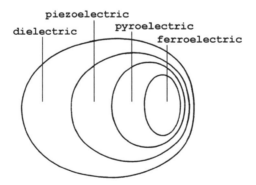

Fig. 4.17   Correlation among piezoelectric, pyroelectric and ferroelectric properties

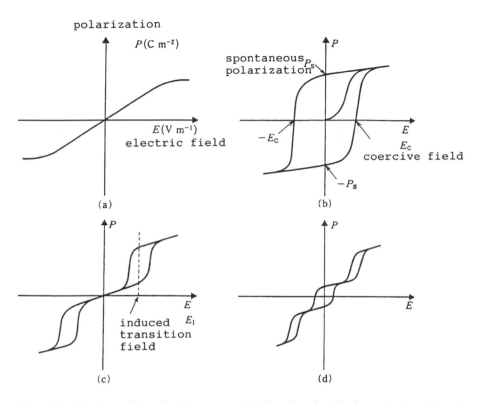

Fig. 4.18 Various dielectric phenomena. (a) Paraelectric; (b) ferroelectric; (c) antiferroelectric; (d) ferrielectric

4.2.3.3 FERROELECTRIC MATERIALS

As discussed above, dielectric materials with spontaneous polarization whose direction can be altered by an external electric field are called ferroelectric. Ferroelectric single crystals have regions similar to magnetic domains of ferromagnetic materials. As indicated in Fig. 4.19, the regions are polarized along the directions of either 90° or 180°. In polycrystalline sintered bodies the direction of polarization of each grain is distributed at random. When a strong DC field is applied to a polycrystalline body with random polarization at temperatures below the Curie point, the polarization of every region is forced to align along the direction of the DC field. This method of aligning the polarization is called poling.

The P–E curve of a ferroelectric material exhibits hysteresis as indicated in Fig. 4.18(b). The magnitude of the dielectric coercive force corresponds to the

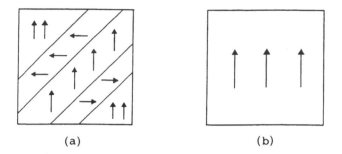

Fig. 4.19 Poling of a ferroelectric body. (a) Before poling; (b) after poling

force required to orient the direction of spontaneous polarization. The characteristics of these non-linear dielectric materials can be assessed numerically by differential and initial dielectric constants.

The temperature dependence of electric susceptibility χ (and dielectric constant ε_r) of an ordinary dielectric material can be derived from Eqs (4.16) and (4.17) and is given by

$$\chi (= \varepsilon_r - 1) = C/T$$

where C is called the Curie constant. On the other hand, ferroelectric materials exhibit their maximum dielectric constants, which become as large as 10 000, near their Curie points (T_c). At temperatures above the Curie point, the polarization is distorted due to thermal vibration and the electric susceptibility obeys the Curie–Weiss law:

$$\chi = C/(T - T_c)$$

Ferroelectric ceramics are listed in Table 4.4. $BaTiO_3$, which is the most technologically important ferroelectric material, has a slightly distorted perovskite structure at temperatures below 120 °C, the temperature at which the cubic to tetragonal phase transformation occurs and at which the positions of oxide ions can be shifted by an external electric field. Thus, $BaTiO_3$ exhibits ferroelectricity. The atomic positions of cubic $BaTiO_3$ is shown in Fig. 4.20. The spontaneous polarization and dielectric constant vary significantly among four polymorphs of $BaTiO_3$. The directions of spontaneous polarization of the tetragonal, orthorhombic and rhombohedral phases are the $\langle 001 \rangle$, $\langle 101 \rangle$ and $\langle 111 \rangle$ directions of the cubic phase. The temperature dependencies of spontaneous polarization and dielectric constant are shown in Fig. 4.21. The figure also indicates that the dielectric constant of the cubic phase obeys the Curie–Weiss law.

An antiferroelectric crystal which exhibits a hysteresis loop as shown by Fig. 4.18(c) has the most stable configuration of dipole moments when they are

Table 4.4 Crystal structures and Curie constants of ferroelectric oxides

Oxide	Crystal structure	Curie point T_c (K)	Curie const. (10^4 K)	Oxide	Crystal structure	Curie point T_c (K)	Curie const. (10^4 K)
$SrTiO_3$	Perovskite	~0	7.0	$LiNbO_3$	Ilmenite	1470	
$BaTiO_3$	Perovskite	373	12.0	$LiTaO_3$	Ilmenite	890	
$PbTiO_3$	Perovskite	763	15.4	$Cd_2Nb_2O_7$	Pyrochlore	185	7.0
$CdTiO_3$	Perovskite	55, 1223	4.5	$PbNb_2O_6$	Tungsten	843	30.0
$KNbO_3$	Perovskite	712	27.0	$Bi_4Ti_3O_{12}$	bronze		
$Pb(Ni_{1/3} Nb_{2/3})O_3$	Perovskite	153			Layered structure	948	
$Pb(Mg_{1/3} Nb_{2/3})O_3$	Perovskite	265		Bi_2WO_6	Layered structure	1223	
$Cd(Mg_{1/3} Nb_{2/3})O_3$	Perovskite	578					

oriented antiparallcl to each other with no external field or when they are parallel to each other with very high external field. In ferrielectric crystals dipole moments cancel each other when no external field is applied, but they are oriented strongly when a high electric field is applied.

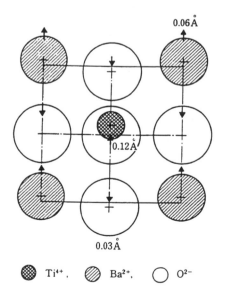

Fig. 4.20 Ionic movements of BaTiO₃ after the application of an electric field. (From G. Shirane, F. Jona and R. Pepinsky, *Proc* ©1955, IEEE, **42**, 1738)

4.2.3.4 CAPACITORS

When a voltage, V, is applied to a plate of dielectric material with thickness d, area A and dielectric constant ε_s, electric charge $Q(C)$ is stored in the plate, which can be given by

$$Q = DA = \varepsilon_0 \varepsilon_s EA = (\varepsilon_0 \varepsilon_s A/d)V \qquad (4.21)$$

Thus the capacitance, C, is given by

$$C = Q/V = \varepsilon_0 \varepsilon_s A/d$$

Dielectric ceramics for capacitors have to satisfy the following two criteria:

(1) A high dielectric constant which does not depend on temperature, voltage, or frequency, and
(2) A high dielectric breakdown voltage and low dielectric loss.

Since no dielectric material can satisfy these two criteria completely, various dielectric ceramics are chosen for a variety of applications.

Ceramic capacitors are classified into the following three application categories: high dielectric constant, temperature compensation, and high

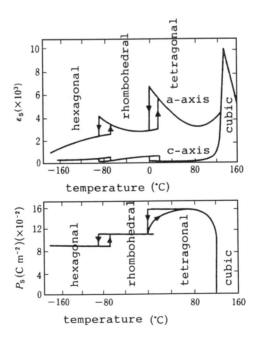

Fig. 4.21 Temperature dependence of a dielectric constant and spontaneous polarization of BaTiO₃. (From W. J. Merz, *Phys. Rev.*, **76**, 1221, 1949)

frequency (microwave). Dielectric ceramics based on $BaTiO_3$ have been widely used for high dielectric constant applications. The minor substitution of Ba with Sr and Pb and Ti with Sn and Zr takes advantage of the high dielectric constant near the Curie point at room temperature. These additives are called shifters. In order to minimize the temperature coefficient of the dielectric constant, Ba is substituted with a small amount of Mg and Ca, which are called depressors.

For the manufacture of multilayer capacitors it is essential for dielectric ceramics to be fired at comparatively low temperatures and to have low-temperature coefficients. These requirements are satisfied by the following lead-based complex perovskite dielectrics:

$$Pb(A_{1/3}^{2+}B_{2/3}^{5+})O_3, \ Pb(A_{1/2}^{2+}C_{1/2}^{6+})O_3$$

where A represents Fe, Mg and Ni, B represents Nb and Ta and C represents W. These dielectric ceramics with broad temperature-dependent dielectric characteristics are called relaxors.

Dielectric ceramics for applications in temperature compensation are required to have very small temperature coefficients and are made of (Sr, Mg, Ca)TiO_3, $SrZrO_3$ and TiO_2. Dielectric ceramics for high-frequency applications are used extensively for wireless telephones and satellite communications which operate at frequencies around a few GHz. Since the dielectric ceramics for high-frequency applications must have low dielectric loss and low-temperature coefficients at high frequencies, they are made of ordinary dielectric and antiferroelectric materials such as Mg_2TiO_4–$CaTiO_3$, BaO–Ln_2O_3–TiO_2 (Ln_2O_3: rare earth oxides), and complex perovskites containing Ta, and (Pb, Ca)ZrO_3. Dielectric characteristics of various dielectric ceramics for capacitor applications are listed in Table 4.5.

Multilayer and semiconductor ceramic capacitors have been employed extensively for applications which require very large capacitance. As shown in Fig. 4.22, multilayer capacitors are made of multiple layers of dielectric ceramics which are sandwiched with layers of metallic electrodes and co-fired. The capacitance of multilayer capacitors increases almost linearly with an increasing number of ceramic layers. In order to prevent the reaction of dielectric ceramics with electrode metals, multilayer capacitors are fired at low temperatures as well as in atmospheres with reduced oxygen partial pressure. The majority of multilayer capacitors are made of complex perovskite dielectrics.

Semiconductor ceramic capacitors are produced by making the surfaces and interfaces of semiconductor ceramics, such as $BaTiO_3$ and $SrTiO_3$, electrically insulating. Those semiconductor ceramic capacitors with insulating interfaces are called boundary layer capacitors. Because of the very low thickness of the dielectric layers (d in Eq. (4.21)), boundary layer capacitors have very high

Table 4.5 Capacitor dielectrics and their characteristics

Application	Material	ε_s	tanδ	Characteristics
Temperature Compensation	MgTiO$_3$–CaTiO$_3$, BaTi$_4$O$_9$	20–100	<0.1%	Capacitance temp. coeff. <±30 ppm/K
	La$_2$O$_3$–TiO$_2$–MgO			
	Nd$_2$O$_3$–TiO$_3$–BaO, (Pb, Ca)ZrO$_3$	20–100	<0.1%	Same as above
Microwave application	Ba(Mg$_{1/3}$Ta$_{2/3}$O$_3$)			2–10 GHz
High dielectric application	BaTiO$_3$ + (NiSnO$_3$, MgO–TiO$_2$) (Pb, La) (Zr, Ti)O$_3$	2000–3000	<2.5%	Wide temp. application
	BaTiO$_3$ + (CaZrO$_3$, CaSnO$_3$, SrTiO$_3$, BaZrO$_3$)	5000–10 000	<5%	General
	Pb(Fe$_{2/3}$ W$_{1/3}$)O$_3$–Pb(Fe$_{1/2}$ Nb$_{1/3}$)O$_3$–Pb(Zn$_{1/3}$Nb$_{2/3}$)O$_3$	10 000–20 000	<5%	Small capacitance temp. coeff. Low temp. sintering
	BaTiO$_3$, SrTiO$_3$ BL	40 000–100 000	<5%	Boundary layer capacitor

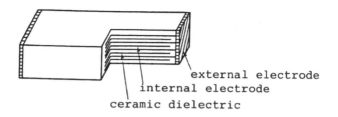

Fig. 4.22   Structure of a multilayer capacitor

capacitance. In order to make the interfaces electrically insulating, metal oxides with a low melting point such as CuO, MnO and Bi_2O_3 are coated on the surface of the capacitors and fired at high temperatures to allow the diffusion of these oxides along grain boundaries. A structural model and equivalent circuit of a boundary layer capacitor are shown in Fig. 4.23.

4.2.3.5   PIEZOELECTRICITY

Dielectric ceramics which exhibit polarization or generate an electric voltage by the conversion of polarization under either stress or strain are called piezoelectric ceramics and are used extensively to convert a mechanical energy (stress T and strain S) to an electric energy (electric field E and flux density D) and vice versa.

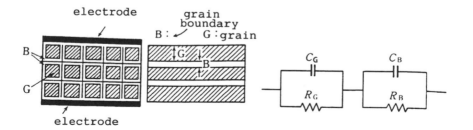

Fig. 4.23   Structural model and equivalent circuit of a boundary layer capacitor. (From R. M. Glaister, Proc ©1961 IEEE 109 B. Suppl. **22**, 3634)

When the polarization of strain is generated parallel to the direction of an electric field or a stress, the piezoelectric equations which correlate these quantities are given by

$$S = s^E T + d_{33} E$$
$$D = d_{33} T + \varepsilon^T E$$

where S^E, ε^T and d_{33} are the elastic compliance (the inverse of Young's modulus) at a constant electric field E, the dielectric constant at a constant stress T and the piezolectric coefficient (CN^{-1}), respectively. d_{33} represents a strain caused by a unit electric field without stress. When a stress is applied, an electric field is induced to maintain $D = 0$ by cancelling out the polarization. The magnitude of the electric field is given by

$$E = -(d_{33}/\varepsilon^T)T$$

On the other hand, when an electric field is applied, a stress is induced to maintain $S = 0$ and thus its value is given by

$$T = -(d_{33}/S^E)E$$

Here, strictly speaking, the values of ε^T and s^E in these two equations are altered by the applied stress and electric field.

The electromechanical combination coefficient k is employed to quantify the effectiveness for the conversion between electric and mechanical energies. From the ratio of the applied electric energy ($\varepsilon^T E^2/2$) to the induced mechanical energy ($s^E T^2/2$) the following relationship can be derived:

$$k_{33}^2 = d_{33}^2/\varepsilon^T s^E$$

Many ferroelectric crystals also exhibit excellent piezoelectric coefficient. As-fired sintered bodies are heated at temperatures above the Curie point. An electric field above its dielectric coercive force is applied during cooling to make them piezoelectric

Today's piezoelectric ceramics are mainly made of solid solutions between $PbZrO_3$–$PbTiO_3$ and are called PZT. $PbTiO_3$ is a ferroelectric material with a large dielectric coercive force and $PbZrO_3$ is an antiferroelectric material. As indicated in Fig. 4.24, the phase boundary between the rhombohedral and tetragonal phases exists at a composition near $Zr/Ti = 53/47$. The electromechanical combination and piezoelectric coefficients are found to exhibit very high values and its applications have been pursued vigorously (Fig. 4.25). Since this phase boundary is almost independent of temperature, its high piezoelectric constant which is about a factor of two larger than $BaTiO_3$ can be applied stably. Furthermore, its piezoelectric characteristics are controlled by the addition of Nb, Ta and La for enhancing the electromechanical

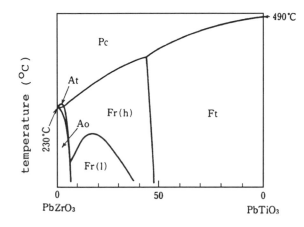

Fig. 4.24 PbZrO₃–PbTiO₃ phase diagram. F—Ferroelectric phase; A—antiferro-electric phase; P—paraelectric phase; c—cubic phase; t—tetragonal phase; r—rhombohedral phase; o—orthorhombic phase; (h)—high-temperature phase; (l) — low-temperature phase. (From E. Sawaguchi, *J. Phys. Soc. Japan*, **8**, 615, 1953: reproduced by permission of The Physical Society of Japan)

combination coefficient, and of Cr, Fe and Mn for enhancing the dielectric coercive force.

A ternary phase diagram among PbZrO₃, PbTiO₃ and Pb(Mg₁/₃Nb₂/₃)O₃, which has a different crystal phase, is shown in Fig. 4.26. Even in this system the piezoelectric coefficient exhibits a maximum near the phase boundary. It is

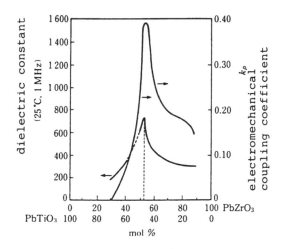

Fig. 4.25 Dielectric constant and electromechanical coupling coefficient of PZT ceramics. (From B. Jaffe *et al.*, *J. Appl. Phys.*, **25**, 809, 1954: © Springer-Verlag)

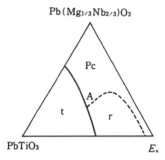

Fig 4.26 PbTiO$_3$–PbZrO$_3$–Pb(Mg$_{1/3}$Nb$_{2/3}$)O$_3$ phase diagram. The letters correspond to those in Fig. 4.24. (From H. Ouchi, K. Nagano and S. Hayakawa, *J. Am. Ceram. Soc.*, **49**, 577, 1965. Reproduced by permission of The American Ceramic Society)

possible to obtain superior piezoelectric characteristics by the combination of ternary components. Pb(Ni$_{1/3}$Nb$_{2/3}$)O$_3$ and Pb(Mn$_{1/3}$Nb$_{2/3}$)O$_3$ are also employed as the third component.

Piezoelectric ceramics have been commercialized in a wide variety of applications. They include (1) ultrasonic vibration and microdeformation elements by the conversion of electric energy to mechanical energy, (2) ignitors and accelerometers by the conversion of mechanical energy to electric energy, (3) piezoelectric transformers and microphones by mutual conversion of electric and mechanical energies and (4) piezoelectric filters for the selection of a narrow range of frequency by the resonance of its intrinsic frequency.

4.2.4 Magnetic Properties

4.2.4.1 ORIGIN OF MAGNETIC PROPERTIES AND MAGNETIZATION

Magnetic properties stem from the magnetic moments of atoms, ions and electrons. In atoms and ions with not completely filled electron orbits the orbital-angular-momentum number l and spin-angular-momentum number s do not cancel out completely and thus the remnant contribute to the formation of magnetic moments. The synthesis of l and s of individual electron is dictated by Hund's rule. The rule says that (1) the spin-angular-momentum number s of each electron takes a value which maximizes the total spin-angular-momentum $S = \Sigma s_i$, (2) the orbital-angular-momentum number l of each electron takes a value which maximizes the total orbital-angular-momentum $L = \Sigma l_i$ and (3) S and L form a combined momentum of $J = |L - S|$ when the d or f orbital is less than half-filled and of $J = L + S$ when they are more than half-filled.

Let us consider Cr^{3+} (d^3) as an example. The d orbital can have five values of l, namely 2 to -2 and the spin S can have two values, namely $\pm 1/2$. Thus the d

Table 4.6 Spin orientation of 3d ions

Ion	Ti⁴⁺ Ca²⁺	Ti³⁺ V⁴⁺	V³⁺	V²⁺ Cr³⁺	Mn³⁺ Cr²⁺	Mn²⁺ Fe³⁺	Fe²⁺	Co²⁺	Ni²⁺	Cu²⁺	Zn²⁺
ml \ n	0	1	2	3	4	5	6	7	8	9	10
2	—	↓	↓	↓	↓	↓	↓	↓	↓	↓	↑↓
1	—	—	↓	↓	↓	↓	↓	↓	↓	↑↓	↑↓
0	—	—	—	↓	↓	↓	↓	↓	↑↓	↑↓	↑↓
−1	—	—	—	—	↓	↓	↓	↑↓	↑↓	↑↓	↑↓
−2	—	—	—	—	—	↓	↑↓	↑↓	↑↓	↑↓	↑↓
S	0	1/2	2/2	3/2	4/2	5/2	4/2	3/2	2/2	1/2	0
L	0	2	3	3	2	0	2	3	3	2	0
J	0	3/2	2	3/2	0	5/2	8/2	9/2	8/2	5/2	0

orbital can accommodate ten electrons. According to Hund's rule, three d electrons of Cr^{3+} have (l,s) values of $(2, 1/2)$, $(1, 1/2)$ and $(0, 1/2)$. Here $S = 3/2$, $L = 3$ and $J = 3 - 3/2 = 3/2$. This type of magnetic moment is called the magnetic moment of a magnetic atom. The spin configuration of ions with a $3d$ orbital is listed in Table 4.6.

The effective magnetic moment, μ_{eff}, of a magnetic atom is given by

$$\mu_{eff} = g\mu_B[J(J+1)] \tag{4.22}$$

where μ_B is the unit of magnetic moment called the Bohr magneton and g is the constant called the g factor.

In $3d$ transition metal ions the L component of orbital-angular-momentum cannot be observed due to the influence of coordinated anions. Thus the overall magnetic moment consists of the spin component only and can be approximated by

$$\mu_{eff} = 2\mu_B S$$

When a magnetic body with isotropic and homogeneous relative magnetic permeability μ_r (dimensionless) is placed in a magnetic field H (A m^{-1}), the magnetic flux density B (Wb m^{-2}) in the magnetic body can be given by

$$B = \mu_0\mu_r H \tag{4.23}$$

where the magnetic permeability of vacuum μ_0 is $4\pi \times 10^{-7}$ Wb A^{-1} m^{-1}. B can also be given by

$$B = \mu_o(H + M) \tag{4.24}$$

where M is the magnetization of a substance and indicates the magnetic moment per unit volume. Since the magnitude of magnetization depends on the magnetic field, the ratio between them χ_m $(= M/H)$, is called the magnetization coefficient. χ_m is also given by

$$\chi_m = \mu_r - 1$$

4.2.4.2 CLASSES OF MAGNETIC PROPERTIES AND MAGNETIC MATERIALS

The magnetic properties of a material are dictated by the magnitude, direction and orientation of its magnetic moments and their correlation. As indicated in Fig. 4.27, materials with randomly oriented magnetic moments are paramagnetic. Materials with parallel magnetic moments without the presence of an external magnetic field (spontaneous magnetization) are ferromagnetic. Those with an antiparallel orientation of equal magnetic moments, thus without overall magnetization, are antiferromagnetic. Materials with antiparallel orientation of unequal magnetic moments and thus with spontaneous magnetization are ferrimagnetic and those which consist of atoms and ions without magnetic moments are diamagnetic.

The application of an external magnetic field to a paramagnetic material causes the orientation of magnetic moments and thus magnetization occurs. For a weak magnetic field, the magnetization is proportional to the external magnetic field. The magnetization coefficient is a function of temperature and decreases with increasing temperature due to thermal vibration. The temperature dependence of the magnetization coefficient can be given by

$$\chi_m = (N\mu_{\text{eff}}^2/3kT) = C_p/T \qquad (4.25)$$

which is called the Curie–Weiss law and is shown schematically in Fig. 4.28(a). In Eq. (4.25) N is the number of magnetic atoms in a unit volume.

Small regions of a magnetic body such as a ferromagnetic body, where magnetic moments are oriented parallel, are called the magnetic domain and their boundaries are called the domain wall. A magnetic body prior to magnetization has a random orientation of magnetic moments and thus has no magnetization. With an increasing external magnetic field, the magnetization increases and approaches a constant value. This phenomenon occurs due to the growth of magnetic domains which have a direction of magnetization very

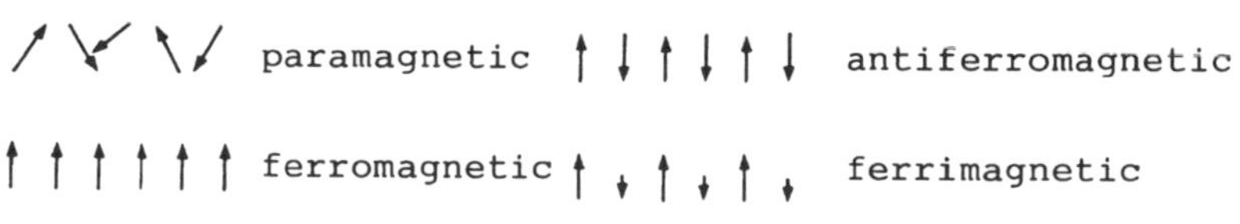

Fig 4.27 Orientation of magnetic moments in various magnetic materials

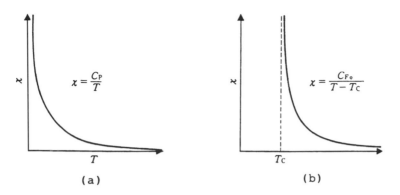

Fig. 4.28 Temperature dependence of magnetization of (a) paramagnetic (excluding metals and (b) ferromagnetic materials

close to that of the external magnetic field through the movement of magnetic domain walls and their rotation. As a consequence, the direction of magnetization is aligned with the direction of the external magnetic field. This orientation will not be eliminated by the removal of the external magnetic field. In order to reverse the orientation, it is necessary to apply a reverse magnetic field whose magnitude is larger than that of the magnetic coercive force. Thus a M–H curve of a ferromagnetic material exhibits a hysteresis as indicated in Fig. 4.29.

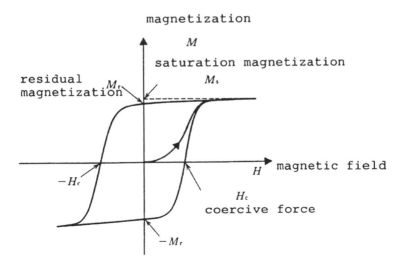

Fig. 4.29 Hysteresis curve of a ferromagnetic material

The spontaneous polarization of a ferromagnetic body disappears above a critical temperature because thermal energies become larger than the mutual interaction force. This critical temperature is called the Curie point, T_c. Above the Curie point, a ferromagnetic material is paramagnetic and its magnetization coefficient can be expressed by

$$\chi_m = C_{FO}/(T - T_c) \ (T > T_c)$$

Magnetic oxides exhibit super-exchange reactions among magnetic ions due to the coupling of orbitals between magnetic and oxide ions. These reactions cause the orientation of magnetic moments, which exhibit ferrimagnetic and antiferromagnetic characteristics.

In ferrimagnetic bodies of iron containing oxides with the spinel structure Fe^{3+} ions occupy tetrahedral sites ($8a$) and Fe^{3+} and M^{2+} (M = Fe, Mn, Co, and Ni) occupy one half of octahedral sites ($16d$) each. Because of the interaction among these ions, the magnetic moments of ions in tetrahedral and octahedral sites are oriented antiparallel. Thus, these oxides are ferrimagnetic. The total magnetic moment is given by the difference in magnetic moments between these two positions and can be calculated simply once the numbers of atoms in these positions are determined. The calculated and measured values of the effective magnetic moments of typical inverse spinels are listed in Table 4.7. The deviations noted in the table indicate that these ferrites do not have a perfect inverse spinel structure. Solid solutions between a ferrimagnetic ferrite and an antiferromagnetic ferrite such as Zn ferrite have effective magnetic moments larger than those of monolithic ferrites. When x mol% of $ZnFe_2O_4$ is dissolved in $Fe^{2+}Fe_2^{3+}O_4$, the composition of the solid solutions can be given by

$$(Zn_x^{2+}Fe_{1-x}^{3+})[Fe_{1-x}^{2+}Fe_{1+x}^{3+}]O_4$$

Table 4.7 Effective magnetic moments of various ferrites (units in Bohr magnetons)

Ferrite	Ion distribution (tetrahedral) [octahedral]	μ_{eff} of tetrahedral sites	μ_{eff} of octa- hedral sites	μ_{eff}/mol Calcu- lated	μ_{eff}/mol Meas- ured
$MnFe_2O_4$	(Fe^{3+}) $[Mn^{2+}, Fe^{3+}]O_4$	5	5+5	5	4.6
Fe_3O_4	(Fe^{3+}) $[Fe^{2+}, Fe^{3+}]O_4$	5	4+5	4	4.1
$CoFe_2O_4$	(Fe^{3+}) $[Co^{2+}, Fe^{3+}]O_4$	5	3+5	3	3.7
$NiFe_2O_4$	(Fe^{3+}) $[Ni^{2+}, Fe^{3+}]O_4$	5	2+5	2	2.3
$CuFe_2O_4$	(Fe^{3+}) $[Cu^{2+}, Fe^{3+}]O_4$	5	1+5	1	1.3
$MgFe_2O_4$	(Fe^{3+}) $[Mg^{2+}, Fe^{3+}]O_4$	5	0+5	0	1.1
$Li_{1/2}Fe_{5/12}O_4$	(Fe^{3+}) $[Li_{1/2}^+, Fe_{3/2}^{3+}]O_4$	5	0+7.5	2.5	2.6

where () and [] indicate the tetrahedral and octahedral sites, respectively. The effective magnetic moment is given by

$$\mu_{\text{eff}}/\mu_{\text{B}} = 4(1 - x) + 5(1 + x) - 5(1 - x) = 4 + 6x$$

As indicated in Fig. 2.20, the measured moment is 5.8 at $x = 0.5$ compared to the theoretical moment of 7. The capability to improve magnetic properties by forming solid solutions with Zn ferrites which cannot be employed alone is another unique feature of ceramic magnetic materials.

Ferrimagnetic materials behave like ferromagnetic materials at temperatures below the Curie point. At temperatures above this point their magnetization coefficients can be represented by

$$\chi_{\text{m}} = (AT - B)/(T^2 - T_{\text{c}}^2) \tag{4.26}$$

Thus the apparent temperature dependence looks very similar to that of a ferromagnetic material.

Crystals which exhibit ferrimagnetic characteristics have the spinel, or the magnetoplumbite or the garnet structure. Ferrites can be distinguished as either soft or hard depending on the difficulty in reversing the direction of magnetization. Hard magnetic materials which are used to fabricate magnets must have a large magnetic coercive force as well as large angular shape ratios ($M_{\text{r}}/M_{\text{s}}$ or $B_{\text{r}}/B_{\text{s}}$). In order to satisfy these requirements, it is necessary to employ magnetic materials with crystal structures which exhibit a large magnetic anisotropy and which prevent the growth and rotation of magnetic domains. Thus Ba ferrite ($BaO \cdot 6Fe_2O_3$) and oxides with the magnetoplumbite structure are the preferred choice for this type of application (see Section 2.3.4). Soft magnetic materials which are used to make high-speed logic operators and magnetic cores of high-frequency transformers with small loss must have a large residual manetization, a small magnetic anisotropy and high electrical resistivity at high frequencies. For this type of application Mn–Zn ferrites with a small amount of CaO and SiO_2 which make grain boundaries highly resistant are the preferred choice.

Magnetic recording materials such as magnetic tapes are essentially assemblies of very fine magnetic particles. A minute part of a magnetic tape is magnetized by a magnetic signal from a magnetic head and data are stored in a form of the magnetization direction. This type of magnetic material must have a magnetic coercive force high enough to maintain magnetization but sufficiently low to allow the erasure of the stored data. In addition, the magnetic materials must be thin films or very fine particles with a particle size of less than a micron. The magnetic materials for this type of application are needle-shaped crystals of γ-Fe_2O_3, CrO_2 and Co-coated iron oxide. Today it is possible to record a data point in an area as small as 1–$10 \, \mu m^2$. Furthermore, optomagnetic recording materials which allow the optical activation and detection of magnetization are currently under development.

4.2.5 Ionic Conduction

Electric conductors with ions as charge carriers are called ionic conductors. Ionic conductors which have high ionic conductivity at temperatures much lower than their melting points are called solid electrolytes. The ionic conductivity of a solid is proportional to the diffusion coefficient of ions (see Section 3.3) and, in general, is higher at higher temperatures. Since the ionic conductivity depends strongly on conductive ionic species and crystal structure, the number of ionic conductors with sufficiently high ionic conductivity suitable for electrochemical applications is quite limited.

Electric conductors with high ionic conductivity are classified into the following three groups:

(1) Ionic conduction by point defects such as vacancies. A good example of this type of ionic conductor is CaO-stabilized zirconia.
(2) Ionic conduction by mobile ions in a crystal structure which has a large number of mobile ionic sites than that of mobile ions themselves. Examples of this type of ionic conductor are α-AgI and $RbAg_4I_5$.
(3) Ionic conduction by the conductive paths of mobile ions. Examples of this type of ionic conductor are $K_{1.6}Mg_{0.8}Ti_{7.2}O_{16}$ with one-dimensional conductive paths, β-Al_2O_3 with two-dimensional paths and Na_3Zr_2P-Si_2O_{12} with three-dimensional paths.

The ionic conductivities of typical ionic conductors are shown in Fig. 4.30.

4.2.5.1 IONIC CONDUCTION BY LATTICE DEFECTS (STABILIZED ZIRCONIA)

Pure zirconia exhibits the following polymorphic phase transformations;

$$\text{Monoclinic} \xleftrightarrow{1000^\circ C} \text{tetragonal} \xleftrightarrow{2370^\circ C} \text{cubic}$$

Since the phase transformation between the monoclinic and tetragonal phases is accompanied with a volume change of about 9%, it is difficult to obtain a stable sintered body. The tetragonal phase of ZrO_2 has a distorted fluorite structure. This distortion is due to the smaller ionic radius of Zr^{4+} (0.82 Å) than the ideal ionic radius of octahedral sites in the fluorite structure (1.02 Å). Thus it is possible to stabilize the tetragonal fluorite structure at a wide range of temperatures by substituting the Zr^{4+} sites by Ca^{2+} (1.03 Å) or Y^{3+} (0.96 Å). These doped ZrO_2 are called stabilized zirconia.

The composition of ZrO_2 doped with x mol% of CaO or Y_2O_3 can be given by either $Ca_xZr_{1-x}O_{2-x}$ or $Y_{2x}Zr_{1-2x}O_{2-x}$ and thus a large number of oxygen vacancies are created (see Section 2.6.4). At high temperatures, oxide ions become mobile via oxygen vacancies. When the concentration of oxygen vacancies is comparatively small, the ionic conductivity is proportional to the

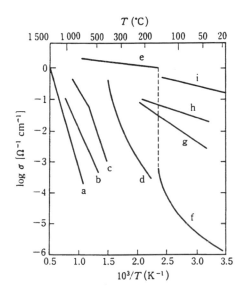

Fig 4.30 Electrical conductivities of major ionic conductors. a—$ZRO_2(+CaO)$; b—$ZrO_2(+Y_2O_3)$; c—$Bi_2O_3(+Y_2O_3)$; d—$AgBr$; e—α-AgI; f—β-AgI; g—$Na_3Zr_2PSi_2O_{12}$; h—Na-β-Al_2O_3; i—$RbAg_4I_5$

concentration of oxygen vacancies. The ionic conductivity saturates at higher concentrations of the dopant and decreases with a further increase in the dopant concentration. At 1000 °C the maximum ionic conductivity occurs for ZrO_2 doped with about 13% of CaO and with about 8% of Y_2O_3. The decrease in ionic conductivity above these concentrations of the dopants is believed to be caused by the formation of associated defects between oxygen vacancies and dopant cations.

The oxygen sensor is one of the important applications of oxide ion conductors. When an electrochemical cell is formed with the stabilized zirconia as electrolyte as shown in Fig. 4.31, oxide ions migrate in the stabilized zirconia from a side with high oxygen partial pressure $P_{O2}(C)$ to a side with low oxygen partial pressure $P_{O2}(A)$:

$$P_{O_2}(C): Pt \parallel \text{stabilized zirconia} \parallel Pt: P_{O_2}(A) \quad P_{O_2}(C) > P_{O_2}(A)$$

As a consequence, positive charges accumulate on the side with high oxygen partial pressure and negative charges on the side with low oxygen partial pressure. Thus an electromotive force is generated between these two sides and is given by

$$E = (RT/4F)\ln\{P_{O_2}(C)/P_{O_2}(A)\}$$

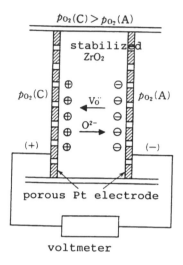

Fig. 4.31 Schematic of a stabilized zirconia oxygen sensor

where R, F and T are the gas constant, the Faraday constant and the absolute temperature, respectively.

Therefore, when the oxygen partial pressure of one side is known, it is possible to determine the oxygen partial pressure of the other side. Through this type of oxygen concentration cells it is possible to determine oxygen partial pressures as low as about 10^{-20} atm. at 1000 °C. These oxygen sensors are employed for measurement of oxygen concentrations in automobile exhausts and molten metals and for the determination of the thermodynamic quantities of various oxides. It is also possible to pump oxygen electrochemically by applying an external voltage to the zirconia cell. This type of pump is called a chemical pump and is used to control the concentration of oxygen partial pressure. Furthermore, when air or oxygen is blown on one side and a gaseous fuel such as H_2 and CH_4 on the other, an electromotive force is generated. When a resistor is connected, an electric current flows between these electrodes. This type of cell is called a high-temperature solid oxide fuel cell and has been developed extensively for the direct conversion of a chemical energy to electrical energy.

4.3.5.2 IONIC CONDUCTION BY THE AVERAGE STRUCTURE

As shown in Fig. 4.32, silver ion conductors such as $RbAg_4I_5$ have an average structure which is very similar to that of α-AgI. In α-AgI a large number of energetically equivalent sites for Ag^+ ions are distributed three-dimensionally among interstices formed by large I^- ions. Thus Ag^+ ions can migrate very

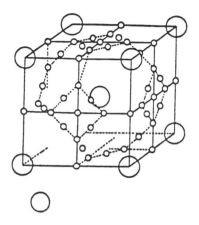

Fig. 4.32   Crystal structure of α-AgI. $\bigcirc$ potential AG^+ site

easily among these sites and a silver ion conductivity as high as that of an aqueous silver solution can be observed even at room temperature. Applications of these silver ion conductors have been sought for electro-potential memory elements and timer switches.

4.2.5.3   IONIC CONDUCTION BY CONDUCTIVE PATHS (β-Al$_2$O$_3$ AND Na$_3$Zr$_2$PSi$_2$O$_{12}$)

β-Al$_2$O$_3$ is an ionic conductor with two-dimensional (namely, layered) conductive paths and its chemical composition is represented by A$_2$O·11Al$_2$O$_3$ where A is the alkali metal ion. As shown in Fig. 4.33, Na-β-Al$_2$O$_3$ has a layered structure which consists of Na$^+$ planes perpendicular to the c-axis and oxygen layers similar to the spinel structure (MgAl$_2$O$_4$) and thus are called spinel blocks. Since Na$^+$ ions can migrate along the plane easily, Na-β-Al$_2$O$_3$ has a very high conductivity of Na$^+$ ions. But because of the anisotropy in the Na$^+$ conduction, the ionic conductivity of polycrystalline Na-β-Al$_2$O$_3$ is about 1/10 that of a single crystal. β''-Al$_2$O$_3$ with a chemical composition of A$_2$O·5-6Al$_2$O$_3$ has a crystal structure similar to β-Al$_2$O$_3$ except that its c lattice constant is one and one half times that of β-Al$_2$O$_3$ and has higher ionic conductivity than β-Al$_2$O$_3$. β''-Al$_2$O$_3$, which is not stable at temperatures above 1500 °C, is stabilized by the substitution of some Al^{3+} ions with Mg^{2+} and Li$^+$ ions. The alkali metal ions in β''-Al$_2$O$_3$ can be substituted with other cations which also exhibit ionic conductivity. The magnitude of the ionic conductivity depends on the ionic radius of the substituted cations.

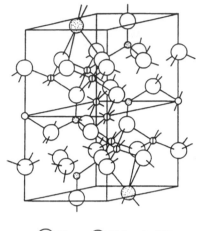

$\bigcirc O^{2-}$, $\bigcirc$ Na$^+$, $\circ$ Al^{3+}

Fig. 4.33 Crystal structure of β-Al$_2$O$_3$

The Na/S battery is one of the major applications of β-Al$_2$O$_3$ and its cell structure is given by

$$\text{Na} \mid \beta\text{-Al}_2\text{O}_3 \mid \text{Na}_2\text{S}_x, \text{S}, \text{C}$$

This battery operates at temperatures above the melting point of Na$_2$S$_x$ which is 285 °C. Because of its high cell voltage and stored electricity per unit weight and long life, the Na/S battery has been developed as a secondary battery for load leveling, which is used to store excess electricity generated during nights and weekends and to discharge at the time of peak electricity demand.

As shown in Fig. 4.34, Na$_3$Zr$_2$PSi$_2$O$_{12}$ has a three-dimensional network structure which consists of tetrahedra of SiO$_4$ and PO$_4$ and octahedra of ZrO$_6$ and has large voids at its center. Na ions can migrate very easily in these voids and thus the ionic conductivity of Na$_3$Zr$_2$PSi$_2$O$_{12}$ is comparable to that of β-Al$_2$O$_3$. These ionic conductors with three-dimensional conductive paths do not exhibit the localization of conductive paths observed in β-Al$_2$O$_3$ and thus are immune to degradation.

4.2.6 Superconductors

The conduction mechanism of superconductors is completely different from that of ordinary electrical conductors. In the latter electrons and electron holes are scattered by the collision among charge carriers and by phonons. On the other hand, all charge carriers in superconductors move in phase and thus do not experience any collision or scattering. Since there is no transfer of

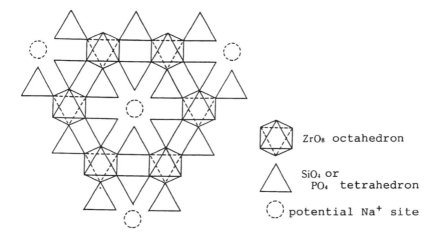

Fig. 4.34 Crystal structure of $Na_3Zr_2PSi_2O_{12}$

momentum among charge carriers, superconductors have no electrical resistance and a current flows in superconductors permanently. Superconductors also exhibit complete antimagnetic characteristics which are called the Meissner effect.

Typical temperature dependencies of electrical resistivity of a superconductor and a metal are shown schematically in Fig. 4.35. Metals have some residual electrical resistivity around 0 K. On the other hand, superconductors have zero electrical resistivity below the superconductive transition temperature T_c.

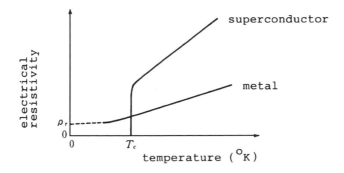

Fig. 4.35 Electrical resistivity–temperature characteristics of a superconductor and a metal ρ_r—residual resistivity of metal; T_c—critical temperature

A group of oxides with a perovskite structure have been found to exhibit superconductive behaviors with high T_c. Examples of the oxides which have a simple perovskite structure are $Ba(Pb, Bi)O_3$ ($T_c < 15 K$) and $(Ba, K)BiO_3$ ($T_c < 30 K$). However, so-called high-temperature superconductive oxides have a layered perovskite structure with layers of CuO_2. From the viewpoint of history and their crystal structures, superconductive oxides can be classified into the following three groups: $(La,M)_2CuO_4$ ($M = Ba,Sr$) ($T_c = 30–40 K$), $YBa_2Cu_3O_7$ ($T_c = 90 K$) and $Bi_2Sr_2Ca_2Cu_3O_{10}$ ($T_c = 115 K$). The crystal structures of $(La, Sr)_2CuO_4$ and $YBa_2Cu_3O_7$ are shown in Fig. 4.36.

$(La,M)_2CuO_4$ was the first oxide found to be superconductive. La_2CuO_4 has the K_2NiF_4 structure which has $Cu–O_6$ octahedra and thus CuO_2 planes. This oxide by itself does not exhibit superconductive behavior, but when some La ions are substituted by M^{2+}, it is believed that the valence of Cu becomes larger than $2+$ (introduction of electron holes) and thus the oxide becomes superconductive. $YBa_2Cu_3O_7$ has the triple-layered perovskite structure from which two oxide ions are missing. There are three pyramidal CuO_2 planes in the unit cell of this crystal structure. $Bi_2Sr_2Ca_2Cu_3O_{10}$ belongs to a group of

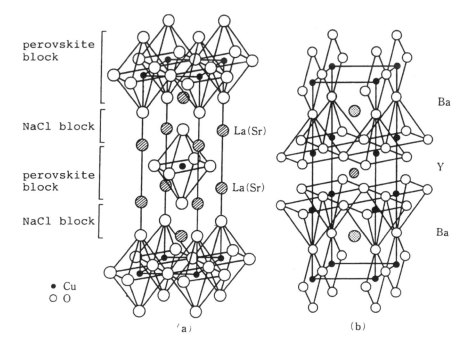

Fig. 4.36 Crystal structures of superconductors. (a) K_2NiF_4 $(La, Sr)_2CuO_4$; (b) oxygen-deficient triple perovskite $YBa_2Cu_3O_7$

superconductive oxides which have the highest T_c known today. Unlike the former two superconductors which have only CuO_2 planes in their crystal structures, the unit cell of this oxide has two pyramidal CuO_2 planes, one simple CuO_2 plane and two layers of Bi_2O_2.

4.3 OPTICAL PROPERTIES

4.3.1 Crystal Symmetry and Refraction

When light is projected onto a material, a part of that light is reflected at the surface and the remaining light will propagate through the material. Unless a light is projected perpendicular to the surface, the light propagating through a material, in general, changes direction. This phenomenon is called refraction. The speed of a light in a material v is smaller than that in vacuum v_0. The ratio between these two speeds $n = v_0/v$ is called the refractive index.

Crystals with a cubic crystal structure and amorphous materials such as glasses have refractive indices which are equal in all directions and are called optically isotropic. A light projected onto a crystal with another crystal structure polarizes into two orthogonal wave components with different speeds. These light waves have their own unique direction of vibration and thus two refracted lights are generated. This phenomenon is called double refraction or birefringence. Those crystals which exhibit double refraction are called optically anisotropic. When an object is viewed through these crystals, double images of the object can be observed.

A light projected parallel to the c-axis of a crystal with either a tetragonal or a hexagonal crystal structure does not undergo double refraction. Those crystals with one direction of crystal axes which do not exhibit double refraction are called monoaxial crystals. On the other hand, crystals with an orthorhombic, monoclinic or triclinic crystal structure have two crystal axes which do not exhibit double refraction and are called biaxial crystals.

A light interacts with a material via the phenomenon of refraction. As a consequence, a variety of optical phenomena have been observed due to scattering, diffraction and interference. Phenomena which are caused by external electric, magnetic and mechanical fields are called the electrooptic, magnetooptic and optoelastic effects, respectively.

In general, the refractive index n is a function of an external field E and is given by

$$n = n + aE + bE^2 + cE^3 + \ldots$$

where a, b and c are constants. When n is proportional to E, the phenomenon is called the primary electrooptic effect whose constant is called the Pockels constant. When n is proportional to E^2, the phenomenon is called the secondary electrooptic effect whose constant is called the Kerr constant. The primary electrooptic effect can be observed in crystals without a center of symmetry such as piezoelectric crystals, but the secondary electrooptic effect can be observed in practically all crystals, including glasses. Typical materials which exhibit the secondary electrooptic effect are $LiNbO_3$, PLZT, K(Nb, Ta)O_3, (Ba, Sr)NbO_3, and $Bi_{12}SiO_{20}$. These materials are used to make optical switches, optical valves to control the amount of lights and optical memories.

The electronic polarization of translucent materials depends on an electric field generated by a light transmitting through a material. When the electronic polarization depends on the terms which are equal to or more than the second power of the optical electric field, the phenomenon is called the non-linear optical effect. Terms which are proportional to the second or third power of the electric field are called either the second- or third-order non-linear optical effects. Crystals with a center of symmetry have non-linear optical effects higher than third order. In general, it is known that ferroelectric materials and piezoelectric crystals with a high index of refraction tend to have a large non-linear effect. The second-order non-linear optical effect which generates highly tuned waves with twice the frequency of the incoming light has been employed to shorten wavelengths of laser lights.

4.3.2 Optical Absorption

The absorption of light occurs because photons can change energy states of electrons. In solids the absorption of ultraviolet light takes place by the excitation of electrons between energy bands, the absorption of light with characteristic wavelengths by impurities and lattice defects, and the absorption of far infrared light by lattice vibration.

Light with energy higher than a band gap is absorbed by the excitation of electrons between energy bands. Since electrons in electrically insulating crystals such as pure SiO_2, which has a band gap of about 8 eV, Al_2O_3 (7.4 eV) and diamond (5.3 eV) cannot be excited by visible light, the light passes through without absorption and thus these crystals are colorless. A semiconductive CdS crystal absorbs light with wavelength shorter than blue, which corresponds to its band gap of 2.4 eV and thus has a color of reddish orange.

When impurity ions are present in a solid, the absorption of light occurs by the excitation of electrons in the impurity between its energy levels. For example, the red color of a ruby crystal is due to the absorption of light with wavelengths of ~ 410 nm and ~ 560 nm by Cr^{3+} ions in Al_2O_3. The coordination of transition metal cations such as Co, Ni, Mn and Fe with anions causes the splitting of their energy levels and thus these cations preferentially absorb light with energies which correspond to energy differences between the energy levels. Alkali halides are colored by heat treatment in Na vapor or by irradiation through X-rays and electron beams. The coloration occurs because the absorption of visible light causes the release of an electron from halogen vacancies with a trapped electron. These lattice defects are called the F center where F denotes color in German (Farbe). Since optical fibers are used to transmit light, which is reflected totally by cores with a refractive index higher than their surroundings, it is necessary to reduce the absorption of light as little as possible. In order to minimize light absorption, the concentrations of impurities and OH groups have been reduced to a minimum by the gas phase method.

4.3.3 Translucent Sintered Bodies

Sintered bodies have complex microstructures which consist of grains, grain boundaries, pores and second phases. As indicated in Fig. 4.37, a light incident to a sintered body experiences a diffuse reflection at the surface and is subsequently absorbed and scattered by inhomogeneities in the sintered body. The intensity of a transmitted light, I, can be given by the following Lambert–Beer rule:

$$I = I_0(1 - R)^2 \exp(-\mu x)$$

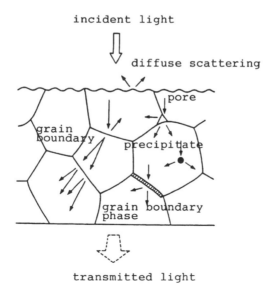

Fig. 4.37   Scattering of light by microstructural imperfections in ceramics

Where I, R, μ and x are the intensity of an incident light, the reflectivity, the absorption coefficient and the thickness of a sintered body, respectively. The reflectivity, R, can be given by

$$R = (n - 1)^2/(n + 1)^2$$

The absorption coefficient, μ, can be given by

$$\mu = \alpha + S_{im} + S_{op}$$

where α, S_{im} and S_{op} are the absorption terms characteristic of electron transition in a sintered body and the scatering terms due to structural imhomogeneities such as pores and second phases and to optical anisotropy, respectively. In order to increase the transmissivity (I/I_0) of a light through a sintered body, it is necessary to reduce μ. Thus it is very important to minimize the scattering term as much as possible. Since α is characteristic of a material, it is a prerequisite to choose a material which does not have a characteristic absorption at desired wavelengths. Namely, in order to make a polycrystalline body transparent, it is necessary to choose a material whose single crystal is transparent.

In order to decrease S_{im}, it is necessary to reduce porosity and densify as much as possible by controlling grain growth during sintering. Translucent ceramics are produced by sintering green ceramics made of high-purity powders at high temperatures (high purity is desirable to prevent the absorption due to solid solution impurities) by adding a small amount of additive as a grain growth inhibitor and by firing in a flowing gas such as H_2 which assists the elimination of pores. Translucent ceramics which have been commercialized so far are listed in Table 4.8 together with the effects of additives and sintering conditions. The hot pressing method has often been employed to make transparent ceramics with high density.

S_{op} is an issue for optically anisotropic crystals which have crystal structures other than the cubic crystal one Al_2O_3 (hexagonal) is an optical monoaxial crystal and thus light tends to be scattered at interfaces such as grain boundaries where refractive indices are discontinuous. As a consequence, transmitted light becomes diffuse. PZT, an important piezoelectric material, is tetragonal and is thus also monoaxial. As indicated in Fig. 4.38, it is possible to bring the ratio between the c and a lattice constants close to unity by dissolving La_2O_3, thus reducing optical anisotropy. Polycrystalline bodies of La-doped PZT have been produced commercially with optical transmissivity between 60% and 70% for light with wavelengths between 0.5 and 6 μm. This type of ceramic is called PLZT (an abbreviation of $(Pb,La)(Zr, Ti)O_3$) and has been used commercially for electrooptic elements such as optical memories. The microstructure of PLZT is shown in Fig. 4.39 and the optical transmissivity curves of various transparent ceramics are shown in Fig. 4.40.

Table 4.8 Sintering additives and conditions of various translucent ceramics

Base oxide	Additives[a]	Optical transmissivity (%)	Wavelength[b] (μm)	Crystal structure	Sintering condition[c]
Al_2O_3	MgO (0.25)	40–60	$0.3 \sim 2$ (1)	h	(1850–1900) × 16 h, H_2
	Y_2O_3 (0.1), MgO (0.05)	70	$0.3 \sim 1.1$ (0.5)	h	1700 × 5 h, H_2
	MgO (0.05)	40	(~ 0.9)	h	1700, -5×10^{-4} mmHg
	Y_2O_3, La_2O_3, ZrO_2 (0.1~0.5), MgO (0.55~1.0)	80[d]	visible light(~ 1)	h	1700, H_2
	MgO (0.05)	85–90[d]	visible light (0.75)	h	(1725–1800) × (17–30) j, H_2
CaO	CaF_2 (0.2–0.6)	40–70	$0.4 \sim 8$ (1.25)	c	(1200–1400) × (0.5–2) h, 5000–8000 psi, 10^{-4}–10^{-6} Torr
MgO	LiF, NaF (1)	80–85	$1 \sim 7$ (5)	c	1000 × 15 min, 1500 psi, *in vacuo*
	NaF (0.25)	Clear	visible light	c	1600 × 111 h, O_2

[a]The numbers in parentheses are wt% of the sintering additives.
[b]The numbers in parentheses are specimen thicknesses in mm.
[c]From left temperature in °C × time, pressure and atmosphere.
[d]Diffuse transmissivity.

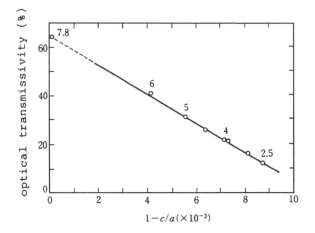

Fig. 4.38 Dependence of optical transmissivity on lattice anisotropy in PLZT. (From K. Miyauchi and G. Toda, *J. Am. Ceram. Soc.*, **58**, 361, 1975. Reproduced by permission of The American Ceramic Society)

4.3.4 Luminescence

When a substance is placed in an electric field or irradiated with light, the substance may radiate light back. The phenomenon consists of the absorption of either electric or optical energy and the radiation of optical energy. A substance which exhibits this phenomenon is called the phosphor and the light radiated is called the luminescence.

Phosphors radiate energies which are generated by the transition between two electron energy levels without losing the absorbed energy in the form of

Fig. 4.39 Microstructure of PLZT. (Photo provided by K. Miyauchi, Hitachi Central Laboratory)

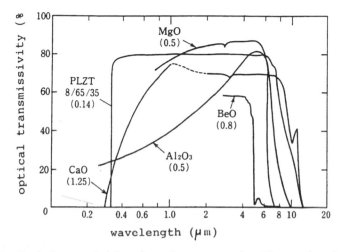

Fig. 4.40 Optical transmissivity of translucent ceramics. The numbers in parentheses indicate specimen thicknesses in millimeters

lattice vibrations among atoms and ions. Thus it is necessary to have a condition where electrons excited by the absorption of an energy are not interfered with by surrounding atoms and ions. Since $4f$ electrons of rare earth elements form an inner electron shell by the shielding effect, the excited electrons are not interfered with by other atoms or ions. Thus rare earth elements are phosphors in themselves. When impurities are shielded from other atoms, the impurities by themselves form luminescence centers. Impurities which form luminescence centers are called activators. ZnS activated by Ag is denoted by ZnS:Ag. The luminescence mechanism of a phosphor with three electron energy levels is shown schematically in Fig. 4.41. Commercial phosphors are listed in Table 4.9.

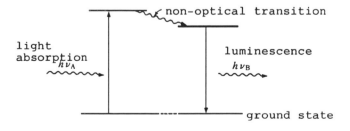

Fig. 4.41 Band model of luminescence

Table 4.9   Applications of various phosphors

Applications	Excitation methods	Phosphors	Color
Color televisions	18–27 kV e-beam	ZnS:Ag, Cl	Blue
		ZnS:Cu, Au, Al	Green
		Y_2O_2S:Eu	Red
CRT	1.5–10 kV e-beam	Zn_2SiO_4:Mn	Green
Electron microscopes	50–3000 kV e-beam	(Zn, Cd)S:Cu,Al	Green
Numerical display	~20 kV e-beam	ZnO	Green
Fluorescent lamps	254 nm UV	$Ca_{10}(PO_4)_6(F,Cl)_2$:Sb, Mn	White
Fluorescent mercury lamps	365 nm UV	$Y(V,P)O_4$:Eu	Red
Copying lamps	254 nm UV	Zn_2SiO_4:Mn	Green
X-ray multipliers	X-ray	$CaWO_4$	Blue/white
		Gd_2O_2S:Tb	Yellow/green
Scintillators	Radiation	NaI:Tl	Blue
EL	$10–5 \times 10$ AC/DC	ZnS:Cu,Mn,Cl	Green
Solid state lasers	Visible light (near UV–near IR)	$Y_3Al_5O_{12}$:Nd (YAG)	IR

From G. Kanou, *Treatise of Ceramic Material Technologies*, p. 465, Industrial Technology Center (1979).

4.4   MECHANICAL PROPERTIES

Unlike metals or polymers, ceramics are hard and heat resistant. Furthermore, it is not possible to form ceramics by plastic deformation. Thus ceramics have the following mechanical characteristics:

(1)   They can be used as high-temperature, high-strength materials.
(2)   Ceramics can be used to machine metals and other ceramics.
(3)   Since the high hardness of ceramics allows precision machining, they can be used as high-precision mechanical parts.
(4)   Ceramics have high strength, but are brittle. Therefore, the amount of mechanical energy required to reach failure is quite small.
(5)   It is difficult to fabricate large ceramic parts.

There are a number of books and publications on the mechanical characteristics of ceramics. Thus no attempts have been made to discuss the mechanical properties of ceramics in detail in this chapter. Instead, the mechanical properties of ceramics will be considered from the crystal chemistry viewpoint.

4.4.1 Strain and Fracture

The mechanical behaviors of a material to fracture can be observed in its stress–strain curve and examples of stress–strain curves are shown in Fig. 4.42. When a strain is small, it is proportional to the stress and becomes zero when the stress is released. Deformation in this range is called elastic deformation. When the stress increases beyond a critical value, the strain increases drastically and cannot be recovered when the stress is released. Deformation which is not recoverable is called plastic deformation.

One of the unique characteristics of ceramics is that, when stressed, they exhibit very small strains and fail before they deform plastically. On the other hand, metals have both large elastic and plastic deformations. At high temperatures, some ceramics also exhibit plastic deformation. Ceramics can be made to withstand large deformation by forming composites.

The energy required to fracture a material is given by the area which is delineated by the stress–strain curve and the strain axis. A material with a large energy of fracture is called tough. Brittleness is a mechanical behavior which can be characterized by a large elastic stress to fracture, but a small total energy to fracture.

A strain, ε, induced by the application of a stress, σ, can be given by

$$\varepsilon = \sigma/E$$

where E is Young's modulus. The volumetric change dV/V induced by the application of an isotropic pressure P can be given by

$$dV/V = P/K$$

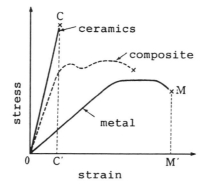

Fig. 4.42 Stress–strain curves of various materials. The fracture energy of a ceramic material is given by the area delineated by O–C–C′–O and that of a metal by O–M–M′–O

where K is the volumetric elastic coefficient. In the elastic range where the deformation of a solid is reversible, these elastic coefficients are important factors which influence mechanical properties such as hardness and strength.

The elastic coefficient denotes a resistance against an external force which is exerted to increase the distance between atoms and ions in a solid. Thus materials which have strong bonds between atoms and ions have a larger elastic coefficient. In addition, they have a short atomic distance and a large packing density of atoms. In general, ionic compounds with a high ionic valence have a large elastic coefficient. In ceramics, carbides and nitrides have large elastic coefficients than oxides. Oxides with a dense crystal structure such as Al_2O_3 and MgO have larger elastic coefficients than oxides with a less dense crystal structure such as SiO_2. Young's moduli of various substances are listed in Table 4.10.

4.4.2 Fracture Strength

The theoretical strength, σ_{th}, of a solid is the maximum stress, σ_{max}, which is required to break a bond between atoms and to create a new surface. The value of the theoretical strength can be given by

$$\sigma_{th} = \sigma_{max} = (\gamma E/a_0)^{1/2} \sim E/10$$

where γ and a_0 are the surface energy per unit surface area and the atomic distance, respectively. Thus, theoretically, substances with large Young's moduli, large surface energies, and small atomic distances should have high strength. Hardnesses and strengths of various substances are listed in Table 4.11.

Measured tensile strengths are much lower that the theoretical strengths discussed above. For example, alumina (Al_2O_3) has a Young's modulus of 390 GPa and its theoretical strength is estimated at about 1/10 of this value. However, the tensile strength of sintered alumina bodies has been measured at about 270 MPa, which is 1/100 to 1/1000 of the theoretical strength. On the other hand, the tensile strength of almost defect-free whiskers is 18 GPa. It is accepted that the reduction of strength is due to the presence of microcracks in

Table 4.10 Young's moduli of various substances

Substance	E (GPa)	Substance	E (GPa)
Diamond (s,p)	1210, 970	PSZ (p)	210
Al_2O_3 (s,p)	460, 390	$MgAl_2O_4$ (p)	240
MgO (s)	250	NaCl (s)	44
SiC (p)	560	Glasses	70–80
Si_3N_4 (p)		Aluminum	60–75

Note: s: single crystal, p: polycrystalline.

Table 4.11 Hardness and tensile strength of various ceramics

Substance	Old Mohr hardness	Ion packing[a]	Atomic bond strength[b]	Tensile strength (MPa)	Young's modulus (GPa)
Calcite CaF_2	4.0	122	0.25	–	–
Feldspar $KAlSi_3O_8$	7.0	115	2.0, 1.5	45	–
Steatite $MgSiO_3$	8.0	16	2.0, 0.75	68	70
Alumina Al_2O_3	9.0	194	1.0	270	390
Silicon carbide SiC	9.2	165	4.0	350	560
Diamond C	10.0	292	4.0	500	1210

[a]Specific gravity/atomic weight of an ion.
[b]Product of electronic charges of a cation and anion/coordination number.
$(10^5 Pa \cong 1\,kg\,cm^{-2})$.

materials. When a microcrack of length $2c$ is present in a material (Fig. 4.43), its strength is given by

$$\sigma_f = (2E\gamma/\pi c)^{1/2}$$

In general, the σ_f of ceramics is about 1/200 of σ_{th}. On the other hand, the σ_f of metals is about 1/3 to 1/50 of σ_{th}. The presence of microcracks is the most critical problem in mechanical applications for ceramics. The seeds for the formation of microcracks are coarse pores, coarse grains, foreign objects and surface grooves.

A stress concentrates at the tip of a crack and stress concentration increases with increasing crack length and decreasing radius of curvature at the crack tip. A fracture occurs due to rapid crack growth under tensile stress. The fracture strength is given by

$$\sigma_f = (1/Y)(2E\gamma_t/c)^{1/2} = K_{IC}/YC^{1/2}$$

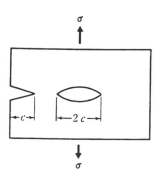

Fig. 4.43 Schematic of external and internal cracks

where Y is the constant which depends on the shape and size of the crack and the loading method, γ_t the effective surface energy and K_{IC} the critical value of stress intensification factor which is called the fracture toughness. K_{IC} can be determined by the fracture stress of a specimen with an artificially introduced crack of known length and has been used as a constant which indicates a material's resistance to fracture. Fracture toughnesses of various ceramics are listed in Table 4.12.

The mechanical strength of ceramics together with K_{IC} is governed by the presence and size of microcracks. In order to use ceramics as structural components, it is vital to fabricate ceramic components which have dense homogeneous microstructures without defects and to detect flaws which will act as seeds for microcracks.

Microstructural factors of ceramics which influence their mechanical strength are porosity and grain size. It is known that strength decreases exponentially with increasing porosity. Several equations have been proposed to represent the relationship between strength and porosity. One of the better known is

$$\sigma_f = \sigma_0 \exp(-bp)$$

where p, b and σ_0 are the porosity, the constant and the strength at $p=0$, respectively. The relationship between the strength and porosity of alumina is shown in Fig. 4.44. The strength also decreases with increasing grain size, which can be represented by

$$\sigma_f \propto d^{-1/2}$$

where d is the grain size. This relationship arises due to the fact that a crack formed in a grain propagates to the grain boundary where the crack is arrested and thus the stress is dispersed through grain boundaries. Ceramic components with coarse grains tend to form cracks along grain boundaries during cooling after sintering and thus have significantly lower strengths.

Table 4.12   Fracture toughness of various ceramics and metals

Material	K_{IC} (MPa m$^{1/2}$)	Material	K_{IC} (MPa m$^{1/2}$)
Al_2O_3	4–4.5	SiC	3.5–6
Al_2O_3–ZrO_2	8–10	Si_3N_4	5–8
ZrO_2	1	BC	5–6
ZrO_2–Y_2O_3 (tetragonal)	6–10	Sialon	5–7
ZrO_2–CaO (precipitation hardened)	8–10	High-strength steel	40–90
ZrO_2–MgO (precipitation hardened)	5–6	Titanium alloy	70

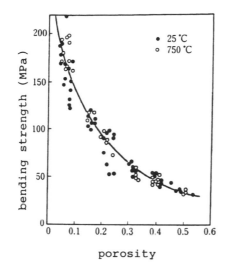

Fig. 4.44   Bending strengths of Al_2O_3 as a function of porosity. (From R. L. Coble and W. D. Kingery, *J. Am. Ceram. Soc.*, **39**, 377, 1956. Reproduced by permisssion of The American Ceramic Society)

When ceramics and glasses are stressed in corrosive environments, they tend to experience slow crack growth below critical stresses. The crack growth rate depends on the stress intensification factor, which is shown schematically in Fig. 4.45. There are three distinct regions in the curve. In region I the crack growth is rate-controlled by stress corrosion at the tip of a crack. In region II the crack growth is rate-controlled by the diffusion of corrodants to the tip of the crack and in region III the mechanical fracture co-exists with the corrosion reaction. As indicated by A in the figure, the rate of crack growth in region I increases with increasing concentration of a corrodant. However, as indicated by B, the rate of crack growth in region II is constant regardless of the material when the concentration of the corrodant remains constant.

4.4.3   High-temperature, High-strength materials

The most noticeable characteristic of ceramics is their resistance to heat and corrosion. Ceramics which exhibit mechanical strength at high temperatures and under corrosive environments can be employed as structural materials. If their strength is sufficiently large, it is also possible to employ them as mechanical materials. Thermodynamically, thermal engines operate more efficiently at higher temperatures. Although thermal engines with conventional metal components cannot operate at temperatures above 1000°C, it is expected

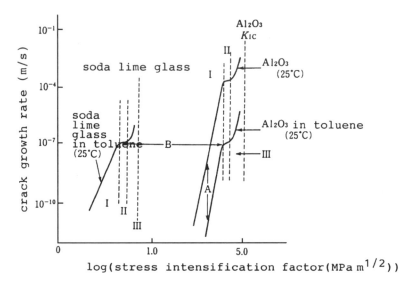

Fig. 4.45 Crack growth rate as a function of stress intensity factor

that gas turbines with ceramic components will be able to operate at temperatures around 1300–1400 °C.

The following characteristics are required for high-temperature, high-strength ceramics;

(1) The ceramics must have a high melting point as well as a high Young's modulus. These ceramics have strong chemical bonds among atoms and thus are inherently strong.
(2) The ceramics must have a small thermal coefficient of expansion but a large thermal conductivity. Ceramics with these characteristics can minimize thermal stresses due to thermal shock.
(3) The ceramics must have high corrosion resistance because they are often used in corrosive environments as in high-temperature steam.
(4) The ceramics must have a large specific strength (strength/density). Ceramics have smaller densities than metals and thus are inherently more advantageous.
(5) The ceramics must have high reliability in strength so that the fatigue failure can be avoided.

Those ceramics which can satisfy the characteristics listed above are mostly covalently bonded ceramics such as Si_3N_4, SiC and Sialon (a solid solution of Si_3N_4–Al_2O_3–AlN). The high-temperature strengths of various ceramics are shown in Fig. 4.46.

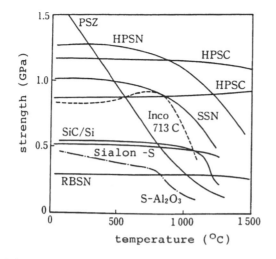

Fig. 4.46 High-temperature strengths of ceramics. SC—SiC; SN—Si$_3$N$_4$; PSZ—partially stabilized zirconia; HP—hot pressed; RB—reaction bonded. (From G. Uegaki, *Ceramic Engines*, Maruzen, 1987, p. 4)

Since these covalently bonded ceramics are difficult to sinter, they are sintered by hot pressing or by reaction bonding using submicron powders with uniform grain size. High-purity submicron powders of SiC are produced by the carbothermic reduction of Si or SiO$_2$ and by the carbonization of halides. High-purity submicron powders of Si$_3$N$_4$ are produced by the direct nitridation of Si and halides and by the thermal decomposition of Si (NH)$_2$.

Sintering additives are also employed to improve the sinterability of these ceramics. For example, MgO and Y$_2$O$_3$ are added to Si$_3$N$_4$. These sintering additives tend to reduce high-temperature strength significantly, a reduction which is caused by the sliding of glassy phases formed along grain boundaries during sintering.

Automobile turbochargers made of Si$_3$N$_4$ have proved to be highly reliable and have been used commercially since 1985.

4.4.4 Toughening

Metals have small Young's moduli but still can absorb large amounts of mechanical energy because they can deform plastically. As a consequence, metals have K_{IC} as large as a few tens of MPa m$^{1/2}$. In some ceramics it is possible to increase K_{IC} by increasing γ_f by the dispersion of fine particles and by stress-induced deformation. In the following section the toughening of partially stabilized and tetragonal zirconias will be discussed as examples.

Zirconia (ZrO_2) with CaO or Y_2O_3 in concentrations which are less than those required to stabilize zirconia completely is called partially stabilized zirconia (PSZ) when the sintered bodies contain a mixture of cubic and tetragonal phases, and tetragonal zirconia (TZP) when the sintered bodies contain only the tetragonal phase. PSZ (CaO 6–8 mol%) is produced first by sintering polycrystalline bodies which essentially consist of only the cubic phase. These bodies are subsequently annealed at 1300–1400 °C, which causes the formation of tetragonal precipitates of about 0.2 μm in size in the cubic grain matrices. TZP (Y_2O_3 2–3 mol%) is produced by sintering polycrystalline bodies with a grain size less than 1 μm at high temperatures where the tetragonal phase is stable. The phase transformation of the tetragonal phase to the monoclinic phase which is the stable phase at low temperatures is diffusionless martensitic and occurs very rapidly. Thus, in general, the tetragonal phase cannot be retained to room temperature. When a small amount of stabilizer is added to zirconia, it is possible to retain fine tetragonal grains to room temperature because of the surface energy effect. The high strength and high toughness of zirconia arise due to the presence of the tetragonal grains in the polycrystalline bodies.

Plausible strengthening mechanisms of zirconia are as follows (Fig. 4.47):

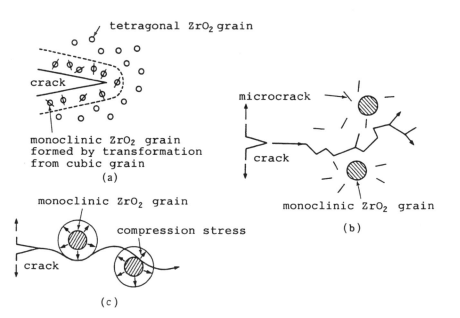

Fig. 4.47 Toughening mechanisms of zirconia ceramics. (a) Stress-induced transformation; (b) crack branching by microcracks; (c) crack deflection by residual stress

(1) Strengthening by stress-induced transformation. The extension of a crack through the metastable tetragonal grain causes stress concentration at the tip of the crack, which promotes transformation to the monoclinic phase. Not only does the transformation absorb fracture energy, it also prevents extension of the crack due to the generation of compressive stress because the monoclinic phase has a larger specific volume than the tetragonal phase.

(2) Strengthening by the formation of microcracks. The volume change associated with phase transformation forms microcracks around the monoclinic grains, which disperse the fracture stress and thus the polycrystalline bodies are strengthened. Microcracks are formed not only by phase transformation but also during fabrication.

(3) Strengthening by residual stress. The monoclinic grains formed by phase transformation are surrounded with localized regions which are under compression. Cracks can pass the regions only with difficulty and thus tend to make detours around the regions. Thus larger amounts of fracture energy can be absorbed.

A composite of Al_2O_3 dispersed with SiC in a grain matrix, which is produced by sintering Al_2O_3 mixed with ultrafine particles of SiC (less than $0.1\,\mu m$ in particle size); exhibits drastic increases in both fracture strength and toughness. It is believed that the composite is strengthened by either mechanism (2) or (3).

Composites with whiskers and fibers also exhibit high toughness. These composites absorb large amounts of fracture energies when whiskers and fibers are pulled away from matrices.

4.5 CHEMICAL PROPERTIES

The chemical properties of solids have been discussed in earlier sections. In the following additional chemical properties will be considered.

4.5.1 Corrosion Resistance

Materials are attacked either by chemical corrosion or by physical erosion. Under attack in various environments, materials change their composition, crystal structure, physical properties and even physical shape. Ceramic materials which are exposed to severe environments such as blast furnace refractories and MHD electrodes must have high corrosion resistance against molten metals and corrosive gases.

Highly corrosion resistant refractories which have been used extensively in various industries can be classified as acidic, neutral or basic refractories. In general, oxides with the composition of RO_2 are called acidic, those with R_2O_3 neutral and oxides with R_2O or RO are called basic. If an oxide which tends to release O^{2-} in a molten salt has high basicity, the dissociation energy ΔH of the oxide such as the reaction enthalpy of $RO = R^{2+} + O^{2-}$ can be a measure of its basicity or acidity. In order to generalize further, the ΔH values are divided by the valence of cations m and by the coordination number of cations around O^{2-} C, and the resulting $\Delta H/mC$ values are listed in Table 4.13, which indicates a good correlation between the values and the conventional classification. Acidic refractories are employed in acidic melts and basic refractories in basic melts.

Gaseous corrosion is another important issue in ceramic materials. In principle, ceramic corrosion by reducing gases such as CO and H_2, by oxidizing gases such as O_2 and by reactive gases such as Cl_2 and SO_2 is governed by thermodynamic and kinetic factors of gas–solid reactions. The corrosivity of ceramics also depends on the microstructural factors such as density, porosity and grain morphology. The corrosivity of various ceramics is compared qualitatively in Table 4.14.

4.5.2 Ion Selectivity

The pH values of aqueous solutions have been measured by glass electrodes which are a kind of sensor for hydrogen ions in the solutions. When ceramics have sensing functions to specific ions, it is possible to use them as ion sensors.

In general, ion sensors are made of ceramics with large ionic mobilities which are called solid electrolytes. For example, the doping of LaF_3 with EuF_2 make

Table 4.13 Relative strength of acidity/basicity of oxides

Oxide	$\Delta H/m$	C	$\Delta H/mC$	
K_2O	277	9	31	↑
Na_2O	307	6	51	Increasing
CaO	833	6	139	basicity
MnO	924	6	154	
MgO	928	6	155	
FeO	940	6	157	
H_2O	511	2	256	
Fe_2O_3	1824	4	456	
Al_2O_3	1824	4	456	
Cr_2O_3	1825	4	456	
B_2O_3	2261	3	754	Increasing
SiO_2	3107	4	777	acidity
P_2O_5	4855	4	1214	↓

From S. Yanagisawa, Y. Watanabe and M. Seki, *J. Jap. Metal Soc.*, **B15**, 478 (1951).

Table 4.14   Corrosion resistance of various ceramics

Ceramics	Acids and acidic gases	Alkali solution and gases	Molten metals
Al_2O_3	Best	Better	Best
MgO	Poor	Best	Best
BeO	Good	Poor	Best
ZrO_2	Better	Best	Best
ThO_2	Poor	Best	Best
TiO_2	Best	Poor	Good
Cr_2O_3	Poor	Poor	Poor
SnO_2	Good	Poor	Poor
SiO_2	Best	Poor	Good
SiC	Best	Good	Good
Si_3N_4	Best	Good	Best
BN	Good	Best	Best
B_4C	Best	Good	–
TiC	Poor	Poor	–
TiN	Good	Good	–

From D. J. Fisher in F. P. Glasser (ed.), *High Temperature Chemistry of Inorganic and Ceramic Materials*, p. 1, Chemical Society, London (1977).

F^- ions mobile and thus it is possible to use the EuF_2-doped LaF_3 as a sensor for F^- ions. Since LaF_3 is difficult to dissolve in water, it can be used as a solid electrolyte in aqueous solutions. Thus, by measuring the electromotive force between a standard solution and an unknown solution, it is possible to determine the concentration of F^- ions in the solution. The electromotive force E can be expressed by

$$E = E_0 + (RT/F)\ln\{C_{F\text{-}s}/C_{F^-}\}$$

where E_0 is the constant for the standard electrode and $C_{F\text{-}s}$ is the concentration of F^- ions in the standard solution.

Substances which are suitable for use as ion electrodes are listed in Table 4.15. These substances hardly dissolve in aqueous solutions and have ionic conductivities at low temperatures.

4.5.3   Ion Exchangeability

Cations such as alkali metal ions which are weakly bonded in crystals can be readily exchanged with other cations. Thus not only is it possible to synthesize various derivatives which have similar crystal structures but different chemical compositions, by exploiting this capability, but it is also possible to synthesize various other compounds with different crystal structures and chemical compositions.

Table 4.15 Compounds for ion electrodes

Ion electrode	Membrane	Usable range (ppm)
Chloride ion electrode	AgCl	0.35–35 000
Bromide ion electrode	AgBr	0.08–80 000
Cadmium ion electrode	CdS	0.005–10 000
Fluoride ion electrode	LaF_3	0.019–19 000
Lead ion electrode	PbS	0.02–20 000
Iodide ion electrode	AgI	0.13–127 000
Silver ion electrode	Ag_2S	0.01–100 000

β-alumina is a well-known solid electrolyte with a chemical composition of $R_2O \cdot Al_2O_3$ (R = Na, K, Ag, Li, Rb and Tl). β_K, β_{Ag} and β_{Tl} aluminas can be obtained by exchanging Na^+ ions in β_{Na}-alumina, which is synthesized from a mixture of sodium carbonate and aluminum hydrate at temperatures above 1000 °C, with other monovalent metal ions in molten salts at temperatures between 300 °C and 800 °C. Li^+ ions in β_{Li}-alumina, which is synthesized by the solid-state reaction between Li_2O and Al_2O_3, occupy Al sites in the spinel block. On the other hand, as indicated in Fig. 4.48, it is possible to replace some Na^+ ions in β_{Na}-alumina with Li^+ ions by ion exchange in molten lithium nitrate. In order to obtain almost 100% β_{Li}-alumina, it is necessary to pretreat β_{Na}-alumina in an aqueous solution of H_2SO_4, which converts β_{Na}-alumina into β_{H3O+}-alumina, and then to conduct the ion exchange in a saturated solution of LiOH.

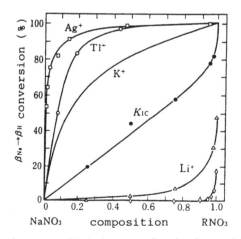

Fig. 4.48 Ion exchange equilibria between β_{Na}-alumina and $NaNO_3$–RNO_3 molten salts. (From Y. F. Yao and J. T. Kummer, *J. Inorg. Chem.*, **29**, 2453, 1967)

The synthesis of $BaTiO_3$ fibers by the reaction of potassium tetratitanate with $Ba(OH)_2$ is an example of the formation of a completely different phase. The crystal structure of potassium tetratitanate consists of TiO_6 octahedra layers and K^+ ions which are located between these layers. Thus potassium tetratitanate has voids in its structure. As a consequence, K^+ ions can be leached out readily into aqueous solutions, which results in a fibrous hydrate. When this hydrate is treated hydrothermally in an autoclave with an aqueous solution of $Ba(OH)_2$, fibrous $BaTiO_3$, can be synthesized.

4.5.4 Inorganic Host Materials

Compounds which have comparatively large voids in their crystal structures can be converted into new compounds by incorporating atoms, or ions or gases in the voids. The base compounds are called the host and those species which are incorporated into the hosts are called the guest. There exist a wide variety of the host compounds with various types of chemical bonds and various sizes of voids, which consist of one-dimensional channels, two-dimensional layers and three-dimensional bulks. These host–guest materials have been developed as catalysts, ionic conductors, molecular sieves and energy-storage materials.

The rutile structure has one-dimensional channel-shaped voids. In this structure, TiO_6 octahedra share corners in the a–b planes. Along the c-axis TiO_6 octahedra share edges and thus form one-dimensional voids where ions with small ionic radius such as Li^+ can be quite mobile. In this system rutile is the host and Li^+ ions are the guest.

Compounds which incorporate atoms, ions and molecules in their two-dimensional layered voids are called either intercalation or layered. Graphite can form a variety of intercalation compounds by incorporating alkali and alkali earth metals, halogen ions, halides and oxides into its crystallographic layers, which are weakly bonded by the van der Waals force. Examples of the intercalation compounds derived from graphite are listed in Table 4.16. These intercalation compounds have been developed as catalysts for polymerization and aromatic alkylation reactions and solid electrolytes for concentration cells.

TiS_2 has the CdI_2 layered structure and forms $Li_xTiS_2(x = 0\text{–}1.0)$ by intercalating Li^+. This intercalation compound exhibits mixed ionic and electronic conductivities and thus it is possible to devise the following oxidation-reduction cell (see Fig. 4.49):

$$TiS_2 \leftrightarrow xLi^+ + TiS_2 + xe'$$

Furthermore, intercalation compounds based on $FeOCl$, MoO_3, TaS_2 and VSe_2 have been the subjects of intensive research and development. Thus it

Table 4.16   Graphitic layered compounds

Starting material	Composition
1. Alkali and alkali earth	
Li	C_6Li; $C_{12}Li$; $C_{18}Li$
Na	$C_{64}Na$
K, Rb, Cs	C_8M; $C_{24}M$; $C_{36}M$; $C_{48}M$; $C_{60}M$
Ca, Sr	C_6M
Ba	C_8Ba
2. Halogen	
Cl_2	C_8Cl
Br_2	C_8Br; C_9Br; $C_{10}Br$; $C_{16}Br$
3. Chloride	C_5CuCl_2; $C_{11}CdCl_2$; $C_{70}WCl_6$, $C_{20}HgCl_2$; $C_{37}UCl_5$
4. Bromide	C_9AlBr_3, Br_2; $C_{18}AlBr_3$; $C_{13}GaBr_3$, $Br_{2.5}$
5. Oxide	$C_{13.6}CrO_3$; $C_{1850}Sb_2O_4$; $C_{100}MoO_3$
6. Sulfide	$C_{432}Cr_2S_{3.5}$; $C_{464}V_2S_{3.5}$; $C_{72}FeS_2$

is expected that a number of new materials will be synthesized in the near future.

Zeolite is the best-known host compound with three-dimensional bulk voids and has a chemical composition of $R_{2/n}O\cdot Al_2O_3\cdot(2+x)SiO_2\cdot yH_2O$. Various classes of zeolite are formed by a combination of AlO_4 and SiO_4 tetrahedra. The fact that the various zeolites have various sizes of channels and bulk voids has been discussed in Section 2.3.6. Some cations migrate quite freely and are trapped eventually in the bulk voids. This ability is called the molecular sieve. In addition, it is also possible to synthesize a variety of catalysts by ion-exchanging them with various cations.

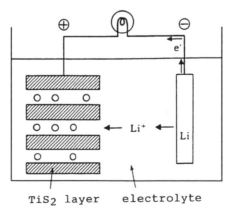

Fig. 4.49   Schematic of a TiS_2/Li battery

Metal hydrides formed by the reaction between metals and hydrogen can be used to store hydrogen. Upon heating, the hydrides decompose and hydrogen gas is released. The hydrides can be used to store energy in the form of hydrogen. In metal hydrides metals are the host and hydrogen is the guest. Numerous metals and alloys such as Nb, Zr, $TiMn_{1.5}$ and $LaNi_5$ have been used as host materials.

As indicated in Fig. 4.50, ReO_3, WO_3, TaF_3 and $TiOF_2$ have fairly open crystal structures, which can be viewed as perovskite structures without cations in A sites and which have three-dimensional channels formed by oxide ion octahedra. Tungsten bronzes with a general chemical formula of M_xWO_3 are formed by incorporating metals such as Na, Li, Ag and Cu in WO_3. These cations have high mobilities in the channels. WO_3 is colorless, but the color of tungsten bronzes changes from yellow to black. By taking advantage of this color change, it is possible to make electrochromic displays (ECD). The oxidation–reduction reaction of the electrochromic display can be given by

$$Li_xWO_3 \leftrightarrow WO_3 \text{ (colorless)} + xLi^+ = xe'$$

An electrochromic display is shown schematically in Fig. 4.51.

Porous glasses are produced by thermal and acid treatments of alkali borosilicate glasses. By controlling the pore size, it is possible to filter gases selectively. In this case the porous glasses are the host and gases are the guest. Attempts have also been made to immobilize enzymes in the pores. As will be discussed in the next section, these porous glasses have attracted much attention as important bioceramics.

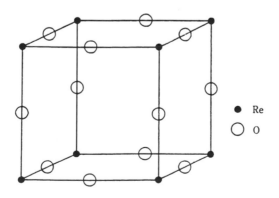

Fig. 4.50 Crystal structure of ReO_3

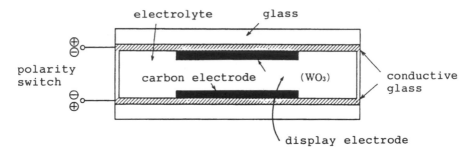

Fig. 4.51   Schematic of an electrochromic display

4.5.5   Biochemical Applications

Because of their chemical stability, durability and high strength, ceramics have been investigated as materials for biomedical applications. Presently they are used as bioceramics in artificial bones, teeth and joints.

Important bodies encountered in the implantation of ceramics in human bodies are poisoning, stability, compatibility with human tissues and various mechanical properties. Examples of bioceramics which have been tested in human bodies are discussed in the following section.

4.5.5.1   ALUMINA

Because of its excellent corrosion and tribological characteristics as well as inertness with human body fluids, alumina has been investigated extensively as a bioceramic. Depending on specific requirements, either single-crystal or polycrystalline alumina can be used. Examples are single-crystal screws for joining bones, artificial teeth and hip joints.

4.5.5.2   APATITE

Human bones and teeth are made mainly of calcium phosphate compounds. In order to duplicate these human parts, attempts have been made to develop polycrystalline hydroxyapatite, $Ca_{10}(PO_4)_6(OH)_2$, which has a similar chemical composition and crystal structure to these human body parts. The growth of new bones in a relatively short period of time has been reported with the implantation of the parts in animals. Long-term strength is a critical deficiency in this class of bioceramics.

4.5.5.3   CARBON

Vitreous carbon, pyrolytic carbon and carbon fibers have been tested in biochemical applications. Although these carbon materials exhibit excellent compatibility and stability in biochemical applications, their use today is very limited due to poor mechanical shock resistance and their black color.

4.5.5.4   BIOGLASSES

Na_2O–CaO–SiO_2–P_2O_5 glasses are called bioglasses. These have excellent compatibility with human bones, but do not have sufficient strength.

INDEX